Further Praise for *Think Like an Engineer*

"In this riveting study... [Madhavan] puts behind-the-scenes geniuses such as Margaret Hutchinson, who designed the first penicillin production plant, centre stage."   *Nature*

"Refreshing, approachable, and highly recommended for anyone interested in understanding engineering... It beautifully conveys the true vision of engineering and its impact on nearly every aspect of life and global progress."

—C. D. Mote Jr., President,

US National Academy of Engineering

"Thoroughly engaging... [Madhavan] leaves little doubt that our world is a better place because of the engineers who inhabit it."

—Henry Petroski, Aleksandar S. Vesic Professor of

Civil Engineering and Professor of History, Duke University;

author of *To Engineer Is Human* and *The Essential Engineer*

"A delightfully interesting book and an essential treatise about ingenuity and systems thinking... Be it a mundane problem or a grand social challenge, the engineering frame of mind offers vital insights and inspiration toward better solutions."

—Rita Colwell, U.S. science envoy,

former director of the National Science Foundation,

and National Medal of Science laureate

D1147485

Guru Madhavan is a biomedical engineer and senior policy adviser. He conducts research at the National Academy of Sciences and has been named a distinguished young scientist by the World Economic Forum. He lives in Washington, DC.

THINK LIKE AN ENGINEER

Guru Madhavan

THINK LIKE AN ENGINEER

Inside the minds that are changing our lives

ONEWORLD

A Oneworld Book

First published in Great Britain and Australia by Oneworld Publications, 2015

This edition published 2016
Reprinted, 2016, 2017, 2018, 2020

ISBN 978-1-78074-864-1
eISBN 978-1-78074-633-8

Printed and bound in Great Britain by Clays Ltd, Elcograf S.p.A.

Oneworld Publications
10 Bloomsbury Street
London WC1B 3SR
England

Stay up to date with the latest books,
special offers, and exclusive content from
Oneworld with our newsletter

Sign up on our website
oneworld-publications.com

MIX
Paper from
responsible sources
FSC® C018072

To my parents and grandparents,
and to the late Chuck Vest—who got me
started.

We taste the spices of Arabia yet never feel
the scorching sun which brings them forth.

—*Dudley North (1602–1677)*

CONTENTS

CONTENTS

THINK LIKE AN ENGINEER

Prologue

INVISIBLE
BRIDGES

(1) NOBODY KNEW where she came from.

It was a celeste-blue afternoon in April 1980, and the world's oldest annual marathon, the Boston Marathon, was on. The streets were flanked by scores of horse-mounted police officers and hundreds of medical responders. A small aircraft skywrote "Fun and Game."

The 26-mile course featured four substantial inclines, but the killer was Heartbreak Hill. A half-mile stretch some 6 miles from the finish line, it usually ended the race for several hundred of the five-thousand-plus runners.

Around half past two, the legendary Bill Rodgers crossed the finish line first for the third straight year, at two hours and twelve minutes. Several minutes later, amid the cheering and hubbub, a young woman in her midtwenties, clad in a white-and-yellow Adidas running suit, sprinted to finish first among women at two hours and thirty-one minutes.

Her name was Rosie Ruiz.

She had set a new record for Boston, becoming the third-fastest woman in the history of marathons. The cheering continued as other lead runners trickled in. One TV journalist immediately announced the time as "a new American record." An interview with Ruiz followed.

REPORTER: What was the time in your first ever marathon, and where was it?

RUIZ: It was two hours, fifty-six minutes, and thirty-three seconds in New York last year.

REPORTER: You improved from two hours and fifty-six minutes to two hours and thirty-one minutes?

RUIZ: I guess so.

REPORTER: What do you attribute that improvement in time to?

RUIZ: I don't know.

REPORTER: Have you been doing a lot of heavy intervals?

RUIZ: Someone else asked me that. I'm not sure what intervals are. What are they?

REPORTER: Intervals are track workouts that are designed to make your speed improve dramatically. If you went from a 2:56 to a 2:31, one would normally expect that you do a lot of speed work. Is someone coaching you or advising you?

RUIZ: No, I advise myself.

REPORTER: It was a fantastic performance, Rosie. Congratulations. Rosie Ruiz, the mystery woman winner.

Race officials were skeptical. Ruiz didn't look tired or sweaty, nor did she possess a marathoner's physique. No one had spotted her at any of the six checkpoints spread over the course. What's more, no one could locate her on the videos at any point during the event—watched by 1.5 million people and covered by more than six hundred reporters.

One eyewitness said, "I just saw someone stumble out of the crowd in front of me, across the street just on Commonwealth Ave., probably about half a mile from the finish. She was in track clothes and wearing a number, but I thought someone had just stumbled into the race; maybe somebody who was a little crazy or something." Several others agreed.

A quick background check revealed that Ruiz was a Cuban immigrant who worked as an administrative assistant in a metal commodities firm in Manhattan. Marathon officials soon found that Ruiz had run only one other marathon before—the 1979 New York City Marathon, a qualifying race for Boston. A news photographer later recalled that Ruiz had cheated in that marathon by taking the subway to Columbus Circle and running from there toward the finish line in Central Park.

Ruiz's breathtaking 1979 fakery led to her "win" at the Boston Marathon, where she ran only the last mile or so. Ruiz maintained her position by looking "as sincere as a nun" and was ready to take a multitude of lie detector tests. After almost a weeklong investigation, the Boston Athletic Association discredited and disqualified Ruiz for cheating her way through the marathon.

She was arrested.

The Ruiz scandal provided plenty of fodder for the media. The U.S. television show *Fridays* poked fun: "Officials became suspicious when she crossed the finish line of the 26.2 mile event wearing open-toed sandals and smoking a cigarette." Someone who personally knew Ruiz told the media, "If you ask her to shed five tears, she'll shed exactly five." Ruiz may well be an ill-famed poster child for marathon offense. As a *New York Times* journalist put it, "Her name endures like a fragile porcelain figure that has been restored after breakage."

The Rosie Ruiz incident presented a conundrum to marathon organizers. Her cheating was obvious, but it exposed the fact that policing a race with thousands of participants is a daunting challenge. How could they prevent future frauds? Engineering provided a solution from a combination of inventions originally fashioned to address two very different issues.

2   IN 1959, railway companies in the United States faced a vexing challenge. The rail system included nearly 1.6 million freight cars, and company officials needed to know the precise location of each car at midnight every day. The whereabouts of these railcars had revenue implications, but there was no way to track them. What was needed was an automated means of identifying and locating the cars.

Around that time, David Collins joined the operations research department of Sylvania, an electric-products company. Collins, who had graduated with a master's degree from the Massachusetts Institute of Technology, loved being an engineer. He enjoyed it so much that he would sometimes joke with his wife that in his next life he wanted to write TV and movie scripts in which engineers were superheroes.

Collins heard about the railway challenge from a colleague. Having worked as a university intern for the Pennsylvania Railroad, Collins knew something about the system. "It fascinated me," he recalled. "I started kicking this project idea around in the lab."

Each railcar was labeled with a horizontal serial number, which was a combination of a six-digit company code and a four-digit car code. Like the cattle brands of western ranchers, these codes came in different reflective colors—red, blue, and white—on a nonreflective black background. They also came in different widths and fonts and had no standard location on the cars. The railcars themselves came in different sizes—tank cars, box cars, and flat cars, the last of which sometimes carried 9-foot-tall semitrailers. These inconsistencies made any attempt to read the railcars' codes difficult. The trains also moved at different speeds—as fast as 60 miles per hour, or at times crawling toward a weigh scale. The need for a dynamic scanning technology that could overcome these problems was obvious.

"So you had a coding system that had been in place for fifty years, but there was no way to acquire that information and make it accessible in machine-readable form," Collins said. He started work on the project in his personal time, eventually gaining his boss's support. Overall, he said, "it was like something that went on at the back of the ranch that nobody really talked about."

Collins's idea was to develop an optical sensor system that could send a white-light beam out to a distant code and decode the signal that reflected back. He focused on primal design elements, such as the code spot size (the specific area where the light is projected and reflected back), scan rate (how many times the code needs to be read each second for an accurate reading), and depth of field (the farthest reading range of the scanner). Initial experiments proved vexing. One of Collins's fellow engineers, Frank Stites, was also baffled by the challenge, but serendipity intervened. Stites inspired Collins to wonder, "Why not turn the code labels sideways?" It was a clever idea.

Scanning the codes vertically—that is, converting a picket-fence arrangement to a ladder-rung style—worked out to be a technically superior alternative. Instead of aiming steady white-light beams at the passing trains—which was usually a hit-or-miss proposition—Collins designed a moving light source with rotating mirrors. His scanner was now able to consistently pick up the patterns from the color codes and decipher a train's information. Then other issues surfaced. Could the scanner work reliably against variable train speeds? Could the sensing be done during snow, rain, and fog? Would the scanning be accurate if there was surface dirt on the codes?

"You couldn't do it in a lab," Collins said. "You were playing with real railcars outdoors, and we didn't own a railroad."

Collins set up a test site near a dedicated train line used to haul materials from New Hampshire to an interstate expansion project in the Boston area. The train usually crossed the site once a day, and he would use that opportunity to test the scanner on hundreds of freight cars with diligence and clockwork precision. He named the scanner KarTrak. Over the following years, Collins boosted KarTrak's performance by replacing white light with helium neon laser. By 1967, its practical application had begun to spread across the railway industry.

The end result? A general-purpose technology that could read codes from a distance.

3 DRIVING ALONG the North Carolina highway one morning in the 1970s in his metallic-green Chevy Nova, George Laurer thought of the times he had hitchhiked with his university buddy along the East Coast after graduation. They were completely broke. They needed jobs. The labor market for engineers in the early 1950s was lukewarm, if not brutal. Laurer was prepared to take any job that would pay him more than $1.50 an hour. Before his job interviews, Laurer would go to the local police station and ask if he could wash up in the bathroom. Months later he landed a plum job at IBM. He spent his entire professional life with the company.

Laurer, now retired, has pale skin, silver-gray hair, and bushy eyebrows. The study in his country home in North Carolina resembles a modern-day Renaissance workshop. It features his collection of mechanical tools, electronic components, technical manuals, and books as diverse as *Complete Book of Welding, Outdoor Projects, American Stamp Album, Basic Bodywork and Painting, TurboCAD Reference Manual,* and *Building Model Airplanes from Scratch*. A balsa aircraft model hangs from the ceiling.

In the early 1970s, inefficiencies in inventory management crippled the grocery industry. Companies needed a way to save money. One idea was to use a code-based system to track grocery products. A committee of top executives from firms such as Heinz, General Foods, Kroger, General Mills, Associated Foods, Fairmont Foods, and Bristol-Myers put out a call for proposals seeking designs for a tracking code. IBM took the challenge in 1971.

Laurer was assigned to the project and instructed by his supervisor to support a "bull's-eye" code developed by another engineer a few years before. "I struggled a day or two with experiments," recalled Laurer. He soon was convinced that the design could not meet the requirements of the grocery industry.

The code had to be no larger than an inch and a half square, and it had to be read easily by both humans and electronics. Next, the symbol had to be printable on oddball-sized products like bar soaps, cereal boxes, and coffee containers. The ten-digit code had to be omnidirectional with an accuracy of at least 99.995 percent, which meant only one error for every twenty thousand items sold. Furthermore, meeting each of these specifications could not add to grocery production costs. Within these rigid constraints, Laurer began to work on a solution.

Risking his job, Laurer went against his supervisor's instructions and set out to create a better approach. The code he designed contained ten vertical black and white stripes with differing widths in a zebra-like pattern. The dark bars absorbed light, and the white ones reflected it back. This reflected light could be picked up by an optical sensor and converted to electrical impulses that were processed by a computer.

For Laurer's prototype demonstration, an ace softball pitcher threw code-labeled beanbag ashtrays as fast as he could over a reader. Each item was read flawlessly. In fact, Laurer's team far exceeded the expectations of the grocery industry: the error rate was as low as one in two hundred thousand. His product was ready. The code selection committee, thrilled with Laurer's creation, called it the Universal Product Code (or UPC). It became industry standard in 1973.

Weeks later, the "golden chicken" problem surfaced. Meat departments in supermarkets had no mechanism to verify that the billed price of the product matched the actual price in the manager's file. Not having a secondary verification meant that occasionally a computer was as likely to undercharge the customer by a few cents as it was to overcharge by a thousand dollars. "Another fact we learned had to do with human nature," Laurer observed. "Most people were willing to forgive the sweet young checkout girl for charging $1.98 for a $1.89 item, but would not forgive a machine for charging $99.99 for a pound of chicken, even though such a gross error would never go undetected. People just do not forgive machines for making errors."

Laurer fixed this problem by adding a price check digit to the Universal Product Code. Over time, additional testing, coupled with improved label printing and detailed receipts, has essentially eliminated these sorts of challenges, thus dramatically changing inventory management and the checkout process.

"It's just a case of sitting down and thinking out every possible solution, step by step, one after another, and also having confidence that there's a solution out there and you can find it," Laurer said. "Not saying, oh well it can't be done."

(4) BORN IN DIFFERENT circumstances and designed independently under different pressures, David Collins's scanner technology and George Laurer's Universal Product Code would eventually converge to create bar codes. This opportunistic combination revolutionized merchandising and founded the modern supply chain system. With the implementation of bar codes, a torrent of new and exciting applications previously imponderable became practical, and which we now take for granted.

From California avocados to Ecuadoran bananas, every perishable good now has an imperishable bar-coded identity.

That is possible because engineers like Collins and Laurer systematically transform problems into opportunities. Their creations were deliberate, disciplined, open-minded, yet grounded in reality. Their process for learning from errors and malfunctions, and then fixing them, was as important as the original idea itself.

Engineers help create *solution spaces*—suites of possibilities that offer new choices, conveniences, and comforts—that redefine our standard of living. With fingerprints on nearly all aspects of the modern world, engineers are engaged in a profession with profound consequences. They are the prospectors of new opportunities, propellers of economies, designers of our material destinies, silent voices in every conversation, and subliminal brokers who facilitate our experiences with the world. Therein lies a paradox: engineering is omnipresent but *invisible*. It tends to be discussed only when an airplane crashes, a bridge buckles, a building crumbles, or a technology fails.

As the *New Yorker*'s John Seabrook has eloquently noted, "Very few inhabitants of modern high-rises know where the load-bearing columns are placed and how they are supported, or whether the building is a frame structure or a tube structure, and almost no one checks above the ceiling tiles to see how the floor overhead is attached to the vertical supports— all decisions that are worked out by the building's structural engineers. The anonymity of the high-rise structural engineer is the reward for his genius. Part of the awe that skyscrapers command lies in their apparent freedom from gravity: they're not just tall; they're effortlessly tall." Look down from an airplane 4 miles above the ground and all you see are systems of nature and engineering. The same is true of the view from the ground, and on into the heavens.

The engineering mind-set resists simple definition. As Craig Barrett, former chairman and CEO of Intel, explains, "It's the rigorous, systematic problem-solving capability that separates engineers from other people who are perhaps more philosophical, argumentative, or blue-sky in their approach to life. I think that's one of the reasons why engineers tend to thrive not only in engineering but also outside of engineering." The engineering mind-set is plug-and-play; it's an all-terrain, multipurpose tool kit. And that's because "engineers are integrators who pull ideas together from multiple streams of knowledge," says Jim Plummer, the former dean of engineering at Stanford University. "They work at the intersection of feasibility, viability, and desirability."

The engineering frame of mind is organic yet synthetic, and engineers are culturally as diverse as what we call "world music." In *Think Like an Engineer*, I'll take you from the standardized controls to the woolly wilderness of the engineering mind-set, using an ensemble cast of engineers. As part of this multidisciplinary excursion, we'll take some scenic routes that demonstrate the power of engineers to convert feelings into finished products. I'll also suggest when thinking like engineers could be a liability. Together, we'll reverse-engineer the engineering mind-set and consider practical aspects that you can apply to your lives.

(5) Soon after the 1980 Boston Marathon, David Collins got a call from the New York Road Runners. The organization was interested in exploring the idea of using bar codes to track runners in its marathons. Collins recalled dismissing that idea instantly: "I said, that's not a good idea. It's just not; forget it." But Fred Lebow, the founder of the New York Marathon, persisted.

Lebow connected with Collins in Boston. They got some people to run around a building in their running suits. Collins tried various ways of using his KarTrak device to scan labels on the runners' bibs. What was possible on freight cars was not possible on people. Humans were unpredictable; they were soaked in sweat, and their labels flapped in all directions. Collins soon hit upon an idea: scan the runners as soon as they reached the finish line and ask them to queue up, since reading them in motion was a hassle. The race officials were still able to determine the finish time for runners with acceptable accuracy.

Collins solved one major challenge of big races—accurately recording time and relative position of the runners—but he couldn't directly solve the Rosie Ruiz problem. It took another creative engineer in the following years and a different technology—radio-frequency identification—to add intelligence to marathons and detect foul play. In recent years, electronic chips have been embedded in a runner's clothes or shoes. Depending on the race, each athlete can be instantaneously tracked, with a resolution of up to a fraction of a second.

"I ran a couple of marathons wearing a bar code because I got so caught up in the customer support issue," Collins offers. "I thought it was a very interesting experience. An 'outlier' application for sure—something I didn't quite predict." On running marathons, "anyone can do it, you know," Collins says. "All it takes is persistence . . . just like engineering."

One

MIXING AND
MATCHING

① Louis XIV loved to have things in order.

As the Sun King of France, he wrote that "good order makes us look assured, and it seems enough to look brave." He built his entire artillery around this principle. But by 1715—at the end of one of the longest regimes in European history and some of its most ruinous wars—Louis XIV's orderly system of defense had become a hodgepodge of improvised work-arounds. His successor, Louis XV, issued a royal ordinance in 1732 that put his lieutenant general, Florent-Jean de Vallière, to work.

Vallière's assignment was to reorganize the artillery, and he was an absolutist. He wanted to create "a system of control: rationality made to serve despotism," as historian Ken Alder writes. Vallière's plans produced a level of centralization previously unimaginable in the French army. Among the impressive results of this exercise were pieces like the Canon de 24—long, thick-cast, bronze-bodied, 24-pound guns that, along with other munitions, were standardized and beautifully ornamented. These guns had superior ranges and were very effective.

But the cannons had one major drawback. Although they excelled at coastal and fortress defense, as well as during siege fights, they fared poorly in offensive warfare. The Vallière cannons were unwieldy and hard to move. Maneuvering them during an open field battle invited a logistical disaster.

In the 1600s, according to one military historian, it took up to twenty horses and an artillery crew of thirty-five men to drag the barrel of a 34-pounder. Even Vallière's 4-pound guns spanned 88 inches and weighed on the order of 1,150 pounds—approximately 288 *times* the weight of the cannonball. The French finally realized that their siege equipment was useful only against static targets. The tactical choices for that era needed to be redefined.

Agility was crucial; swiftness, essential. They needed a new system.

(2) As a CHILD, Jean-Baptiste Vaquette de Gribeauval was curious about instruments. Born in 1715 into a family of lawyers, he later went to an artillery school to learn ballistics engineering. At seventeen, he volunteered for the French army. In 1748, Gribeauval modified the design of a naval carriage unit to potentially transport guns for offensive operations. In 1749 he was promoted to captain. Later that year, Vallière rejected Gribeauval's proposal for mass-producing his gun carriage units, which may have made it easier to move the bulky cannons.

Gribeauval was frustrated. He respected the rule-bound order of the Vallière cannons but felt they were mired in an artisanal mode of production. There was something even more demoralizing for Gribeauval: he lacked authority within the corps. His ideas carried no influence. With jealousy and rivalry the norms of the day, military promotions occurred at a sluggish pace. Overall, he had little incentive to stay in his position.

The French and Prussians had been allies since 1741, but the 1756 signing of the first Treaty of Versailles between France and Austria—two staunch rivals—infuriated Prussia. The Franco-Prussian relationship corroded. Prussia quickly formed an alliance with Great Britain and attacked France and its partners—Austria, Bavaria, Russia, Saxony, and Sweden—thus kindling the Seven Years' War, what Winston Churchill in later years referred to as "the first world war."

After the war began, Austria realized that it desperately

needed good military engineers; its armies had an overhang of poorly trained technical officers who went up the ranks through favoritism, not merit. Gribeauval saw an opportunity and got himself lent to Austria as a wartime ally. He intuitively knew that lightweight cannons were critical for offensive warfare, something that the Vallière system sorely lacked compared to the mobile armies of Prussia. Gribeauval employed to great technical success some of his redesigned guns and his much-improved 1748 naval carriage unit.

Following this demonstration, Gribeauval steadily gained influence in the Austrian army. His eyes were now set on reforming the Austrian manufacturing process and taking it beyond artisanship. He inspired his superiors by noting that Austria had massive advantages over the French cannons. "An enlightened man without passion who understood the [relevant] details and had sufficient credit to cut straight to the truth, would find in these two artilleries the means to compose a single one which would win almost every battle in the field," Gribeauval wrote. "But ignorance, vanity, and jealousy always intervene; it is the devil's work and cannot be changed as easily as a suit of clothes, it costs too much and one runs a great danger if one is not sure of success."

In 1762, at the peak of the Seven Years' War, Gribeauval made his move. During the siege of Schweidnitz, he commanded a handful of troops against a massive Prussian army unit. Gribeauval held out against the Prussians for sixty-three days in one of the bloodiest battles of that time, consuming about three thousand lives. Even his enemy, Frederick the Great, was impressed by Gribeauval's methods. In the end, though, the Prussians won. Gribeauval was arrested but released at the end of the Seven Years' War.

Gribeauval was now an "authentic military hero." The French, having observed Gribeauval's rise, now offered him authority and a lucrative return package. In an audacious first step, Gribeauval dethroned Vallière's system—which in his mind had contributed to the French defeat. As a result, a great rivalry rose within the corps. Gribeauval and Vallière were embroiled in "the 'Star Wars' dispute" of the day, writes Alder, on "a public debate over the offensive and defensive capabilities of the nation and effectiveness of high-tech gadgetry." It was a duel between the "ancients" and the "moderns."

Gribeauval began to hone the design of the French cannons. He was obsessed with precision and laid out specifications that could be verified to within one-thousandth of an inch—less than the thickness of a single sheet of paper. Using skilled metallurgists and sophisticated boring machinery, he added elevation screws that enabled precise adjustments offering highly effective aiming. The addition of rear sights to better position the guns, and leather straps to pull them, turned out to be of tremendous help to the soldiers during combat operations. Gribeauval introduced larger wheels for the guns so that they could tread easily on rough terrains, and he replaced wooden axles with cast iron for easy maintenance and repairs. These were small but critical adjustments that improved the cannons' usability. They also defined Gribeauval's tactics.

Vallière's cannons had to be returned to gunsmiths for maintenance and troubleshooting. In contrast, Gribeauval's designs could be readily dismantled and reconfigured. One part of the cannon could be used to replace a different part with the same specification. This interchangeability was possible because of the principles of "parameter variation," in which the various components are tested individually while others are kept constant, similar to how algebraic equations are solved. The "combination of factors"—which Gribeauval began to glean from his artillery examiner Pierre-Simon Laplace, a mathematical genius—was experimentally applied to maximize output, Alder explains.

In this process, Gribeauval had created a technology development platform for the future. His strategy was to achieve what no one had been able to in the past: high efficiency, uniformity, and exchangeability. Product construction tables were developed, manufacturing standards were introduced, and protocols were established for easy and fast servicing of the guns. This systematic process led to the development of lightweight cannons and made the Gribeauval system the most effective artillery in Europe.

This was a radical idea in an era of siege warfare. "The most significant innovation" with Gribeauval's system was "that it was indeed a system: a thorough synthesis of organization, technology, material, and tactics," writes historian Howard Rosen. "Every aspect of the system, from the harnessing of the horses to the selection and organization of personnel, embodied a single functional concept. *Utility* was its principle, *mobility* was its goal."

None of these relied on the classical rules of the day.

(3) THE CORE OF the engineering mind-set is what I call *modular systems thinking*. It's not a singular talent, but a mélange of techniques and principles. Systems-level thinking is more than just being systematic; rather, it's about the understanding that in the ebb and flow of life, nothing is stationary and everything is linked. The relationships among the modules of a system give rise to a whole that cannot be understood by analyzing its constituent parts.

A specific technique in modular systems thinking, for example, includes a functional blend of *deconstructionism* (breaking down a larger system into its modules) and *reconstructionism* (putting those modules back together). The focus is on identifying the strong and weak links—how the modules work, don't work, or could potentially work—and applying this knowledge to engineer useful outcomes. A related design concept, exploited especially by software engineers, is *stepwise refinement*. Every successive change that engineers make to a product or service expressly contributes to a better result or the development of alternative solutions. Even within this framework of product development, there's a top-down design strategy—"divide and conquer"—in which each subtask is separately attacked in a progression toward achieving the final objective. The opposite of this approach is a bottom-up design in which the modules are recomposed.

Ruth David, a U.S. national security expert and former deputy director for science and technology at the Central Intelligence Agency, frames the issue this way: "Engineering is synonymous not only to systems thinking but also *systems building*. It's the ability to look at a problem in different ways. One not only has to understand the pieces and their interdependencies, but also really understand the totality and what it means." It's one of the reasons the engineering mind-set is portable across many sectors of society and effective both for individuals and for groups.

Modular systems thinking varies with contexts because there is no widely accepted "engineering method." Engineering Dubai's Burj Khalifa is different from coding the Microsoft Office Suite. Whether used to conduct wind tunnel tests on World Cup footballs or to create a missile capable of hitting another missile midflight, engineering works in various ways. Even within a specific industry, techniques can differ. Engineering an artifact like a turbofan engine is different from assembling a megasystem like an aircraft, and by extension, a system of systems, such as the air traffic network.

With the changing realities around us, the nature of engineering is also changing. In addition to being the "hardware of culture," it's reliably an engine for economic growth. In the United States, for instance, recent estimates suggest that less than 4 percent of the total population are engineers who disproportionately help create jobs for the remainder. Certain engineering innovations do supplant human jobs, but they routinely help generate new opportunities and pathways for development.

◆　◆　◆　◆

THE ENGINEERING MIND-SET has three essential properties.

The first is the ability to "see" *structure* where there's none. From haikus to high-rise buildings, our world relies on structures. Just as a talented composer "hears" a sound before it's put down on a score, a good engineer is able to visualize—and produce—structures through a combination of rules, models, and instincts. The engineering mind gravitates to the piece of the iceberg underneath the water rather than its surface. It's not only about what one sees; it's also about the unseen.

A structured systems-level thinking process would consider how the elements of the system are linked in logic, in time, in sequence, and in function—and under what conditions they work and don't work. A historian might apply this sort of structural logic decades *after* something has occurred, but an engineer needs to do this preemptively, whether with the finest details or top-level abstractions. This is one of the main reasons why engineers build models: so that they can have structured conversations based in reality. Critically, envisioning a structure involves having the wisdom to know when a structure is valuable, and when it isn't.

As can be seen from the works of Vallière and Gribeauval,

military systems are renowned for their structured approach to innovation. Consider, for example, the following catechism by George Heilmeier—a former director of the U.S. Defense Advanced Research Projects Agency (DARPA), who also engineered the liquid crystal displays (LCDs) that are part of modern-day visual technologies. His approach to innovation is to employ a checklist-like template suitable for a project with well-defined goals and customers.

- *What are you trying to do? Articulate your objectives using absolutely no jargon.*

- *How is it done today, and what are the limits of current practice?*

- *What's new in your approach and why do you think it will be successful?*

- *Who cares? If you're successful, what difference will it make?*

- *What are the risks and the payoffs?*

- *How much will it cost? How long will it take?*

- *What are the preliminary and final "exams" to check for success?*

At a basic level, this type of structure helps ask the right questions in a logical way.

The second attribute of the engineering mind-set is the

adeptness at designing under *constraints*. Any real-world scenario has constraints that make or break our performance potential. Given the innately practical nature of engineering, the pressures on it are far greater compared to other professions. Constraints—whether natural or human-made—don't permit engineers to wait until all phenomena are fully understood and explained. Engineers are expected to produce the best possible results under the given conditions. Even if there are no constraints, good engineers know how to apply constraints to help achieve their goals. Time constraints on engineers fuel creativity and resourcefulness. Financial constraints and the blatant physical constraints hinging on the laws of nature are also common, coupled with an unpredictable constraint—namely, human behavior.

"Imagine if each new version of the Macintosh Operating System, or of Windows, was in fact a completely new operating system that began from scratch. It would bring personal computing to a halt," Olivier de Weck and his fellow researchers at the Massachusetts Institute of Technology point out. Engineers often augment their software products, incrementally addressing customer preferences and business necessities—which are nothing but constraints. "Changes that look easy at first frequently necessitate other changes, which in turn cause more change. . . . You have to find a way to keep the old thing going while creating something new." The pressures are endless.

The third attribute of the engineering mind-set involves *trade-offs*—the ability to make considered judgments about solutions and alternatives. Engineers make design priorities and allocate resources by ferreting out the weak goals among stronger ones. For an airplane design, a typical trade-off could be to balance the demands of cost, weight, wingspan, and lavatory dimensions within the constraints of the given performance specifications. This type of selection pressure even trickles down to the question of whether passengers like the airplane they're flying in. If constraints are like tightrope walking, then trade-offs are inescapable tugs-of-war among what's available, what's possible, what's desirable, and what the limits are.

Science, philosophy, and religion may well be in the business of pursuing truth as it looks to them, but engineering is at the center of producing utility under constraints. Structure, constraints, and trade-offs are the one-two-three punch of the engineering mind-set. They are to an engineer as time, tempo, and rhythm are to a musician.

4 SEPTEMBER 12, 1962. "If I were to say, my fellow citizens," President Kennedy told a gathering in Houston, Texas on a warm day,

> that we shall send to the moon, 240,000 miles away from the control station in Houston, a giant rocket more than 300 feet tall, the length of this football field, made of new metal alloys, some of which have not yet been invented, capable of standing heat and stresses several times more

than have ever been experienced, fitted together with a precision better than the finest watch, carrying all the equipment needed for propulsion, guidance, control, communications, food and survival, on an untried mission, to an unknown celestial body, and then return it safely to earth, re-entering the atmosphere at speeds of over 25,000 miles per hour, causing heat about half that of the temperature of the sun . . . and do all this, and do it right, and do it first before this decade is out—then we must be bold.

The most crucial part of Kennedy's vision was not the technical ambition but the assertion "before this decade is out." This time pressure forced the project engineers to accomplish the objective. The Apollo 11 mission successfully landed on the moon on July 20, 1969, ahead of the actual deadline. The process leading up to the lunar landing created several valuable by-products, including new materials like carbon fiber and advanced navigation systems that are now used by commercial airlines. Though engineering is what put humans on the moon and brought them back safely, the whole effort is often called rocket "science."

If the core of science is discovery, then the essence of engineering is creation. Going back to the very roots of human history, as a civilization we were tool builders before we were discoverers. In fact, many tools of engineering have enhanced our capabilities to produce better science. Scientists now increasingly rely on engineering to obtain unimaginable amounts of data and results in order to propose, test, or advance their theories. Engineering does rely on natural laws and scientific evidence, but it also helps generate new bodies of scientific knowledge. Airplanes flew before a formal study of aeronautics became a reality. Steam engines gave birth to the science of thermodynamics. Further, the industrial revolution robustly expanded avenues available to scientific inquiry. According to Tom Peters, a professor at Lehigh University in Pennsylvania engineers even "gladly 'creatively misinterpret' scientific method or results if that helps get the job done."

History shows that "most of the Ages are characterized by engineering," reminds Dan Mote, president of the U.S. National Academy of Engineering. "The Stone Age . . . was named after chipping rocks by hand to create tools; the Bronze Age was named for the smelting of tin and copper to cast weapons, tools, and artifacts; the Iron Age was named after hammering and bending iron to create farming implements and tools; and the Silicon Age reflects the material foundation for electronics manufacturing," Mote says. "OK, the Ice Age was not a human creation—as a natural phenomenon it belongs to science."

Scholars have gone on to argue that engineering occupies a separate realm of knowledge and practice—one that's far more secure and reliable than other intellectual traditions rooted in philosophy—and therefore deserves distinct respect. Since Plato, a Western intellectual bias emphasizing the superiority of "pure" knowledge has downplayed engineering. It's also unfortunate that "science and technology" are almost always discussed together without mention of engineering, even though technology is an outcome of both science and engineering. "Science is a tool of engineering, and as no one claims that the chisel creates the sculpture, so no one should claim that science makes the rocket," writes engineering historian Henry Petroski. "Relying on nothing but scientific knowledge to produce an engineering solution is to invite frustration at best and failure at worst."

George Whitesides, an eclectic Harvard chemist-turned-engineer, offers another useful comparison between science and engineering. If science is interested in "tracing a mechanistic pathway from ions and neurotransmitters to the Brahms *Requiem*," then engineering is focused on providing "practical solutions to sequester unlimited amounts of carbon dioxide, and providing unlimited power and clean water with a guaranteed 30 percent after-tax return on investment, using equipment not readily available in Namibia." Knowledge for the sake of knowledge has its role, but practical reality shapes social progress.

Neuroscientist Stuart Firestein likens the scientific process to finding a black cat in a dark room—especially when there's no cat. This is different from the image people typically have of scientists: "patiently piecing together a giant puzzle." Scientific knowledge goes hand in hand with ignorance. Science is driven by a continuous "communal gap in knowledge," as Firestein frames it. The knowledge is not always useful and can't "be used to make a prediction or statement about some thing or event. This is knowledgeable ignorance, perceptive ignorance, insightful ignorance," he adds. "It's not facts and rules. It's black cats in dark rooms."

Mathematician Andrew Wiles, quoted in Firestein's book *Ignorance: How It Drives Science*, amplifies this view: "It's groping and probing and poking, and some bumbling and bungling, and then a switch is discovered, often by accident, and the light is lit, and everyone says 'Oh, wow, so that's how it looks,' and then it's off into the next dark room, looking for the next mysterious black feline."

The value of science, we've been taught, resides in its objectivity. Ideally, science eschews planned outcomes. Engineering frequently runs contrary to this idea: at its finest, it's allied with subjectivity. Yet objectivity can be an especially helpful principle for engineers when trying to prevent or analyze failures. In a real, symbiotic way, science and engineering help each other to uncover their inconsistencies and shortcomings. Unlike the blueprint of the Brooklyn Bridge, for science there is no final draft of knowledge. Our hypotheses can take us in any direction.

(5) BECAUSE I WAS BORN in a lower-middle-class, orthodox Hindu Brahmin family in rural Tamil Nadu, a coastal state in southern India, my pathway to engineering emerged from a sink-or-swim circumstance. It wasn't a chemistry set—my parents couldn't afford it—that sparked my interest in science, nor did I have a congenital instinct

to build Lego robots. Perhaps my earliest technical curiosities blossomed from watching coal-fired steam engines in the early 1980s—thanks to my father, who took me on his morning bicycle rides to the local railway station.

As far as I can remember, I wasn't even particularly good in mathematics. Before I took my exams, I was sure to visit a temple of Ganesha—the elephant god—to pray for good grades. My paternal grandfather was a farmer during the day, and a priest at dawn and dusk. During our early years, my younger brother and I were his assistants in our village temple near Tiruvannamalai—a grouping of hills considered to be older than the Himalayas. We were mesmerized by our grandfather's soulful Sanskrit mantras during his morning and evening prayers. Our bedtime treats were his stories from the ancient epics *Ramayana* and *Mahabharata* as we fell asleep on straw mats.

Throughout my education in India, the vigorous environment drove my aspirations. Sharp focus, fewer distractions, and first-rate academic performance were the most desired outcomes for my schools. In essence, my education was on an assembly line. During my time at school, I mused over what else I might be interested in; degrees in medicine, commerce, and engineering were especially prized in my local culture. I'd soak my feet in the waves of the Bay of Bengal at the Madras marina hoping for inspiration. My father—a chemist-turned-accountant—and my stay-at-home mother encouraged me to pursue any field that held my interest.

The brutal competition at school didn't give me—or my

brother and our friends—the time or liberty to explore, experiment, and "fall in love" with something. Frankly, my entry
into engineering was like an arranged marriage—a pragmatic
route for success where I was trained. I chose to major in
instrumentation and control systems engineering—back then
a fresh, relatively uncrowded, and marvelously challenging
degree program offered by the University of Madras. My
eventual interests in developing biomedical technologies,
coupled with a generous scholarship, flew me to a graduate
school in New York—a month before September 11, 2001.

Over time, I realized engineering was a force larger than
the mathematical models I wrestled with, more meaningful
than the electronic circuits I designed, more precise than the
sensors and devices I tested, more insightful than the software codes I debugged, and far more breathtaking than the
vapid technical jargon could ever convey. What had started
out in me as a synthetic fascination gradually matured into an
organic, renewing love for engineering.

6) GRIBEAUVAL'S TOOLS were also developed within
the triad of structure, constraints, and trade-offs. The
resulting creations offered a blueprint for precision and large-
scale manufacturing that has since affected the far reaches of
our society. Moreover, these ideas helped initiate the age of
mass production, which then spurred on the spread of modern engineering.

With structure, Gribeauval brought a superior sense of

purpose to cannons and their use. He guided the mixing and matching of relevant parts, capitalizing on the practice of interchangeable design, which remains an active approach in the world of engineering. One of the technical aspects underlying interchangeability is the practice of "functional binding." The modules weren't a mishmash bricolage, but a strategically interconnected system that had to fulfill a single function. With this strategy, errors were quickly identified, tested, fixed, and retested—a practice that the future assembly line technology would go on to perfect with precision. The cannons had to perform with accuracy and durability. In the centuries preceding sophisticated modeling and simulation software, engineers like Gribeauval relied on their calculations and tacit know-how to arrive at stable solutions. They usually overengineered temples, bridges, castles, and other systems to ensure that they wouldn't fail.

Constraints were constant companions to Gribeauval. The stakes were monumental—there was a war to be won—and therefore his solutions had to work. For natural philosophers like Galileo and Newton, the study of ballistics had been what Ken Alder calls a "mathematical gymnasium," located purely in the confines of their intellect. Mathematics, for them, was "a form of 'descriptionism,' a way to quantify how changes in certain measurable parameters affected some other relevant parameter," explains Alder. "Mathematics, more often than not, enabled engineers to *evade* real causal explanation." Unlike others, who didn't necessarily have to apply their knowledge to concrete use, Gribeauval had to overcome the real difficulties of wind and air resistance over the course of improving his cannons' projectiles. He capitalized on the method of parameter variation—by deconstructing and reconstructing the cannon modules—to assess the strengths and weaknesses in his manufacturing system and how he could improve the cannons' performance. His guns needed to fire accurately and as expected. That's how they fulfilled their purpose.

Finally, circumstances required Gribeauval to make design choices. Was achieving improved maneuverability more important than designing cannons that fired with increased force? Could excessive weight be reduced without increasing the gun's failure rates? As one design feature, Gribeauval eliminated superfluous decorative ornamentations on the cannons. Agility trumped cosmetics. His judicious trade-offs, coupled with continued experimentation on parameter variation, dramatically improved the ability to produce and transport better artillery.

When serving the Austrian army, Gribeauval was astonished by the prevailing custom of flagrant favoritism in the hiring and promotion of incompetent technical officers. Good engineers really suffered. He wrote,

> *[Engineers] are treated in a way that is harsh and even indecent. . . . When an officer, no matter how junior, is dispatched on some mission, he invariably takes a couple of engineers with him to see to the hard and uncongenial parts of the task; they load the blame on them [if] any-*

thing goes wrong, but take the credit if [it] turns out well. Just look at the state of the engineers . . . you will see that most of them have lost their horses and money, and that they are worn by exhaustion and maltreatment.

To circumvent the problem, Gribeauval helped codify a merit-based training system for his labor force, helping to create an era of what Alder calls "enlightenment engineering." Geometry, technical drawing, and analytical calculus were used to assess principal competencies, and then became standard courses in artillery schools and military academies. Centuries later, these courses remain at the heart of engineering education. In transforming his technical conscience into applied realities, Gribeauval helped maximize defense innovations, create jobs, proliferate new industries, and advance national security. After all, as the saying goes, "In theory, there's no difference between theory and practice. In practice, there is."

Two

OPTIMIZING

1 IN THE EARLY 2000s, Stockholm's traffic congestion was getting out of control.

Commuting times had exploded. Delays and frustrations
were mounting. The Swedish capital's productivity ground
to a halt during rush hours. One obvious way to resolve the
challenge was to boost capacity by building another bridge.
This strategy had worked before—Stockholm already had
dozens of bridges—and, after all, it was the "Venice of the
North." Stockholm city officials instead paused for reflection.
They then hired an unusual group of consultants: engineers
from IBM.

For IBM, the project was more like planning a rescue
mission than preparing a traffic angioplasty to unclog Stockholm's arteries. To tackle the unknown, the IBM crew started

installing sensors around the city to monitor traffic. IBM used 430,000 transponders that accrued data and collected 850,000 photographs. Using this information, IBM produced a total systems model by mathematically analyzing all traffic nodes and seemingly unrelated bottlenecks. The result of this detailed effort convinced Stockholm officials that they shouldn't build any new bridges or roads. Instead, they should start charging commuters who wanted to use existing bridges and highways during peak hours.

The results of congestion pricing were astonishing. During the trial run of this new scheme in 2006, traffic congestion in Stockholm decreased between 20 and 25 percent. The wait time for commuters plummeted by one-third on average— even to almost one-half. Public transportation regained popularity. The scheme helped remove a hundred thousand cars from the road. Levels of carbon and other particulate emissions plunged. In 2007, Stockholm passed a permanent referendum implementing a camera-based toll system. Sweden's successful experiment garnered attention. Cities in Asia, Europe, and North America began to contemplate sophisticated congestion-pricing approaches.

◆ ◆ ◆ ◆

TRAFFIC JAMS are like leaky buckets: with more supply, they only get worse. In addition, the carrying capacities of roads are fixed, so handling extra cars during peak hours presents a nearly insurmountable challenge.

A recent urban mobility report from the Texas A&M Transportation Institute noted that the fifty-six billion pounds of annual carbon emission in big cities of the United States during peak hours is "equivalent to the liftoff weight of over 12,400 Space Shuttles with all fuel tanks full." The nearly three billion gallons of fuel consumed to create those emissions is "enough to fill four New Orleans Superdomes."

On an individual level, these figures become dramatic. The personal cost to the average commuter has more than doubled, as has wasted fuel, over the last thirty years. American commuters, the report noted, "spent an extra 38 hours traveling in 2011, up from 16 hours in 1982." This equates to five lost days spent in traffic. In the U.K. the situation is worse: the average commuter now spends 211 hours travelling per year, the equivalent of more than a month's full-time work.

"Today we have innumerable road sensors and cameras on the ground that automatically upload data so information can be shared and analyzed in nearly real-time," writes Naveen Lamba, the global industry lead for IBM's Intelligent Transportation products. Sensors and transponders that IBM used to support its analyses turned out to be indispensably helpful in mapping traffic flow. "Once data is five to seven minutes old, it's too late to make any changes that will reduce congestion," Lamba adds. "Once a commuter is stuck in gridlock, it's too late to find an alternate route." Forecasting the traffic demand is an added challenge; even real-time data are often insufficient.

Building our way out of traffic congestion is not always a viable option. "We have to learn to get more productivity out of existing assets using technology," Lamba declares. In Stockholm, IBM took a modular approach in trying to understand every piece of the system that may be directly or indirectly contributing to the traffic jam. The outcome was to create a new electronic infrastructure: car tags linked to an electronic or convenience store account for payment. This approach influenced public behavior and the very social process of commuting. The revenue that came from the tolls could be applied to the upkeep of a city's highway system and other activities. Peak pricing, in this case, was not a point solution but a *platform solution* that simultaneously tackled a number of other challenges. A leaky bucket became an ocean of opportunities.

A solution that doesn't work in one setting may be transformational in another. Unlike Stockholm, a village in Africa might benefit from an additional road or a bridge to increase public access to services and opportunities. With a decent road, people who never thought about getting a vehicle might choose to buy one. A road means more mobility, and more mobility means more commerce.

Traffic congestion is a function of human behavior. It comes in the form of *latent preferences* wired in each of us—how we choose to travel from one place to another. As a result, public behavior plays a pivotal role in the success or failure of infrastructure design projects or policies. By and large, that's because traffic, like any social arrangement, is a complex system composed of multiple systems that interact with one another with no main controller. Their collective effects are by nature nonlinear, often leading to unpredictable behavior called *emergence*. Even an infinitesimal change (a single orange traffic cone) can lead to an unpredicted impact (a freeway traffic jam) across a system of systems consisting in part of roads.

On this topic, Vinton Cerf, the coinventor of the Internet, offers an astute perspective. One day Cerf was trying to pour peppercorns into a grinder using a funnel. "A few peppercorns got through and then they got stuck. If I had dropped them in one at a time, there would have been no problem," Cerf observes. "But because I poured several of them into the funnel, the emergent property in this case was congestion."

This basic understanding of the complex, large-scale effects (such as behavior change) arising from simple rules (peak pricing) is a helpful notion for optimization. "The thing is that you can't create congestion with one peppercorn," Cerf adds. "And the most interesting part is that there isn't anything about a peppercorn that will explain to you much about its congestive properties, other than maybe the fact that it's due to *friction*."

◆   ◆   ◆   ◆

ANYONE CAN CLAIM to optimize something in words, but in practice it's a different story. Optimization is akin to attending a gym and committing to repetitions for strength training. How can we get the best results out of a workout in the shortest period? How can we continually make something better?

Optimization has two basic components. The first component is an *objective* focused on maximizing or minimizing an outcome variable that is usually a function of something else. Gribeauval's optimization objective was to inflict maximum losses on his enemy, with a broader goal of winning the war. Optimization also includes a *constraint*, consisting of limitations to which the objective is subjected. Operations researchers who use models and study ways to improve efficiencies would consider Gribeauval's goal a classic "weapon target assignment problem" for which they would develop an algorithm. With limited time and resources, how could Gribeauval find the right set of tools—or combinations of tools—and position them optimally to achieve his objective?

Engineers use a variety of modeling techniques to arrive at approximate representations of reality that, by their nature, are not exact. Models are of two basic types: implicit and explicit. In *implicit models*, "assumptions are hidden, internal consistency is untested, their logical consequences are unknown, and their relation to data is unknown," as Joshua Epstein, a professor at Johns Hopkins University, describes. In this regard, "when you close your eyes and imagine an epidemic spreading, or any other social dynamic, you are running *some* model or other. It is just an implicit model that you haven't written down." In *explicit models*, the assumptions, empirical caveats, and equations are clearly presented for analysis and verification. With one set of assumptions, "*this* sort of thing happens. When you alter the assumptions, *that* is what happens," Epstein adds.

Among the many benefits of modeling are to "demonstrate

trade-offs and suggest efficiencies" Epstein highlights, or even "reveal the apparently simple to be complex, [and the complex to be simple]." Models expose areas where more data are needed and reveal what work needs to be done. Collecting data on road use patterns from all corners of Stockholm strengthened IBM's model and its eventual decision to recommend congestion pricing.

There are no perfect models for optimization. Every model is limited by its assumptions and criticized for reducing reality to simple equations. "Simple models can be invaluable without being 'right,' in an engineering sense," says Epstein. "Indeed, by such lights, all the best models are wrong. But they are fruitfully wrong. They are illuminating abstractions." But the primary purpose of using models to support optimization is to develop a structure that makes constraints and trade-offs clear.

While models are valuable, they also mislead at times. A familiar fallacy among engineers is to assume that a model working well at one level will also function the same way at a different scale. Not necessarily. In fact, emergent properties in complex systems are almost always a function of scaling. Construction engineer John Kuprenas and architect Matthew Frederick derive this insight from the Victorian astronomer Sir Robert Ball:

An imaginary team of engineers sought to build a "super-horse" that would be twice as tall as a normal horse. When they created it, they discovered it to be a troubled, inef-

ficient beast. Not only was it two times the height of a normal horse, it was twice as wide and twice as long, resulting in an overall mass eight times greater than normal. But the cross sectional area of its veins and arteries was only four times that of a normal horse, calling for its heart to work twice as hard. The surface area of its feet was four times that of a normal horse, but each foot had to support twice the weight per unit of surface area compared to a normal horse. Ultimately, the sickly animal had to be put down.

Models are support systems. They are a means to aid decisions; they are not the final decisions themselves. By illuminating the pluses and minuses surrounding the final objective, good models provide a reality check for optimization. In IBM's case, the primary objective was to minimize traffic congestion in Stockholm, which turned out to be a function of automobile usage during peak hours. The constraints included fixed road capacity, the local government's budget, and people's latent preferences. A natural starting point to fully understand and optimize such a complex system was to build a model.

(2) IN THE EARLY 1940s, the U.S. Post Office Department faced a crisis. A large number of postal workers had left the department to serve in the military during the Second World War. Annual mail volume was skyrocketing (it reached forty-five billion pieces by 1950), thanks in large part to the explosive growth in direct-mail advertising over the preceding two decades. How could the department optimize postal delivery around the country?

Pressures relating to cost, efficiency, accuracy, and delivery schedule—and perhaps the institution's future—led the postal department to take an engineering approach. The fascinating results are one of the great strengths of today's U.S. postal system, which have benefited the rest of the world.

The system's designers segmented the United States into "zones," each with a distinct five-digit identification number. And so it was that in 1963, following two decades of research and engineering, the postal service announced its implementation of the Zone Improvement Plan code. The ZIP code established a whole new system for connecting senders and recipients of mail.

In a process emblematic of modular systems thinking, the developers of the ZIP code divided the country into ten sections numbered 0 to 9. They started on the East Coast, assigning Maine the number 0, and moved across the country westward. ZIP codes in New York and some of its neighboring states started with 1; those in the Washington, DC, area began with 2; the West Coast states with 9; and so on. Other numbers in the code further parsed these zones according to central post office hubs and the nearest post office in a particular region.

Specialized machinery was developed to help sort mail for each zone. It took time to refine the accuracy, since the process included a human element. An operator had to key the

ZIP code for each envelope or package into a sorter machine. The keying process led to typos and errors. A letter, for example, that was supposed to go to Chemult, Oregon, could end up being routed to Custer, South Dakota, and subsequently rerouted to the postal hub in Denver, Colorado.

To twenty-first-century ears this system might sound inefficient, but in the 1960s, ZIP codes were "revolutionary because of the idea that the mails were being processed based on a *number* code," says Nancy Pope, a technology historian at the Smithsonian National Postal Museum. The ZIP codes were also helpful in streamlining mail sent to U.S. cities with common names, like Greenville or Salem or Springfield.

Before mechanization, postal staff members were hand-sorting the pieces. In that situation, "even if you're really good at it, you're not going to do more than sixty letters a minute," Pope says. "I mean, that would be topping out as the best sorter on record for the postal service." On average, most workers managed twenty to thirty pieces a minute, and because these processes were manual, errors were possible. With the advent of automation, the game completely changed. Machines were able to process up to two thousand pieces a minute, if not more, and therefore a system like the ZIP code laid the groundwork for improved efficiency throughout the postal enterprise.

Federal buildings—the U.S. Capitol, the White House, and the Pentagon, for example—were assigned their own special ZIP codes. Other countries soon began to adapt the ZIP code idea to create their own versions of numeric or alphanumeric postal index codes, such as the postcode system in the U.K. ZIP codes became an iconic engineering solution—and an integral part of commerce—that made dramatic improvements in the efficiency of the postal service, even as it reduced costs and errors by integrating new postal technologies. The development of the ZIP code was the result of *master planning*—a long-term strategy that's typical of many successful (and unsuccessful) large-scale engineering, architectural, and military projects. Sometimes a deliberate, carefully planned deconstruction is required for the creative reconstruction of a system.

Not everyone was thrilled about the implementation of ZIP codes. People didn't like having to remember five numbers. The three-digit area codes for telephone numbers had also come into effect. And businesses had started to require social security information for income tax. It all appeared to be a numeric conspiracy—even a communist plot. A systems optimization concept like ZIP codes required a massive national campaign to persuade people to embrace it, and that included a cartoon character called Mr. Zip. Ethel Merman, famous for her theme song "There's No Business Like Show Business", lent her brassy voice for a promotional jingle: "Welcome to Zip Code . . . learn it today. Send your mail out . . . the five-digit way."

The impact of ZIP codes extends far beyond postal applications. Online enterprises now routinely capitalize on the twentieth-century postal engineering infrastructure to collect demographic, behavioral, and other information about their customers. These codes have become a requisite feature for megaprojects such as the census, direct-mail campaigns, targeted micromarketing applications—what some praise as "recommender systems" and others critique as "consumer espionage"—and for authorization at gas station pumps and supermarkets. In the U.K., for example, the term "postcode lottery" describes the inequality in the delivery and quality of health care and other public services—the idea that where people live may define the standard of services they can expect to receive.

Engineering, as it should be clear by now, is not only about technology—that is, replacing manual labor with machinery. In equal or more parts, it's about strategy. The development of ZIP codes—similar to how IBM saw a structure in the traffic mess and used it to change public behavior—was a simple yet profound strategy in optimization. It helped solve a practical problem more than a technical one.

Scholars and practitioners have used a variety of terms to argue about the distinction between technical problems and practical problems. Examples include "problems" and "messes"; "tame problems" and wicked problems"; "high grounds" and "swamps"; "hard problems" and "soft problems." These terminologies signal a basic split. In the first case of each example, something well defined needs to be solved. In the second case, the issue being tackled cannot be readily solved using only equations or analytics but requires also an understanding of human and other factors, which often contribute to emergent properties. Both ZIP codes and congestion pricing are examples in which engineers blended technical and social factors in practice.

We'll now see how a major Internet products firm applied this sort of optimization to mapping and cataloguing our world.

3   GOOGLE'S GOAL is ambitious: organizing the world's information. The company's New York City operations are housed in a 1930s-era former Port Authority building in the lower west side of Manhattan. Google's primary-color theme radiates the possibility that it could also be a day care center for grown-ups. Past the clickety-clack of keyboards, and the abundance of free food in the pantries, sits the office of Alfred Spector, vice president of research and special initiatives. He likes to use Google Maps to monitor traffic intensity and plan his trips. "I've only missed the train out of Grand Central to Pelham at most three times in the last six years," Spector says confidently.

Spector and his colleagues operate with the belief that every piece of information has an expiring window of opportunity. The trick is in seizing the data at the right time and in the right context so that the information becomes useful. Near-real-time technologies like Google Maps are governed by the concept of *continuous optimization*. "We get very effective traffic data now in New York with red, dark red, green, yellow indicators; it's quite realistic," Spector says. "We might be reducing peak traffic on New York City roads by directing people to better approaches."

The idea of altering traffic in response to congestion or an incident is not a new challenge. Operations researchers recognize this as a resource reallocation problem that surfaces especially during an emergency: employ the evacuation route for people to disperse efficiently, and use the ingress routes for responders to enter the affected area. Google's innovation has been to direct the power of information to users so that they can make data-aided, adaptive choices.

Spector's colleagues have written that, when trying to build something new, like Google Maps, "instead of debating at length about the best way to do something," they "prefer to get going immediately and then iterate and refine the approach." This is to help support Google's primary mission: "solve really large problems." Consider this fundamental challenge: there are about 50 million miles of paved and unpaved roads in 195 countries. "Driving all these roads once would be equivalent to circumnavigating the globe 1,250 times—even for Google, this type of scale can be daunting," the project engineers write.

Google's engineers began their project by acquiring visual data from around the world—thanks to recent developments in street-level panoramic imagery and user-provided photographs. The next step was to develop a large-scale systems model that "includes detailed knowledge about one-way streets and turn restrictions (such as no right turn or no U-turn)," the engineers explain. Using this information, Google then converted the position of the sensor embedded in its camera— nowadays also in our phones—into accurate road-positioning data through a method called *pose optimization*. The process was driven not by a single algorithm, but by a group of inter-connected tools.

Google engineers branched out to auction algorithms—typically used for determining the best offer on an item with concurrent bidders—to predict the real-time traffic demand for commuters simultaneously interested in taking the same route. They used image-processing techniques for creating "depth maps" to encode 3-D data on distance, orientation, and other local information such as roads, pavements, build-ings, and construction work. They relied on remote sensing and pixel-level analyses of satellite images to help generate multiple views of any location—whether the Eiffel Tower or an abandoned mining town in the Alaskan wilderness. The engineers collectively made the best of these tools—and they continue to do so with several others—to create a better value for the users of Google Maps.

"The idea of driving along every street in the world taking pictures of all the buildings and roadsides seemed outland-ish at first," the engineers add, "but analysis showed that it was within reach of an organized effort at an affordable scale, over a period of years." From Spector's viewpoint, it was a basic question of cost-effectiveness. Google Maps started out as an engineering trade-off concerning efficient logistics—that is, could the mapping be done?—but then was followed up by an economic argument on the potential market for the application.

"It would turn out to be feasible," Spector says.

❖   ❖   ❖   ❖

BEING DATA DRIVEN is a precondition for optimization. This idea has influenced every industrial sector. In the telecommunications industry, for example, in recent years the "volumes that are going over our mobile data networks are up 25,000 percent and they are doubling every year still," notes Randall Stephenson, the CEO of AT&T Corporation. In the airline industry, a Boeing aircraft flying from London to New York pumps out 10 terabytes of operational data every thirty minutes during the trip.

Being data driven is also only one part of optimization. Understanding user needs is another critical component. Consider this scenario from Norman Augustine, retired CEO of Lockheed Martin Corporation: Suppose you did a survey and asked passengers what they would like in a new airplane. Say you found out that they would like to get to their destinations sooner. For an aerodynamics expert, it may become an issue of the airplane needing to fly faster. For a systems engineer the approach is different.

Using modular thinking, a systems engineer would break down the entire travel process into its component parts. Flying in the airplane is one part of the system among many. There's the getting to the airport part, finding a parking spot part, navigating the terminal part, ticketing part, processing the baggage part, waiting for security part, waiting to board part, boarding part, and getting to the final destination part.

All these parts—and several more—contribute to the speed, efficiency, and performance of the entire system. The systems engineer can try to optimize individual parts while paying attention to trade-offs and constraints. With modular thinking, the solutions might come down to how to get through security faster, how to improve the boarding process, and how to retrieve luggage quickly.

Things get more complicated when you add in Mother Nature, the ultimate system of systems. In the case of aviation, for example, weather is a huge, unpredictable factor in optimization. Similarly, in early days, lack of measurements and data forced engineers involved in constructing pipes and sewers to make immediate assumptions and judgments. "If you want to build a tunnel, you have to deal with a geological medium that's constantly changing and interacting with other systems," says Wayne Clough, a geotechnical engineer who is secretary of the Smithsonian Institution, and former president of the Georgia Institute of Technology. "You need to have a strong systems approach that will allow you to adapt to changing environments." Modern technologies enable extraordinary amounts of data collection about nature. But using that information for any form of optimization will always be a challenge.

We can try our best using our technologies, but ultimately Mother Nature wins.

(4) I WENT TO business school while simultaneously working on a doctorate in biomedical engineering. My goal was to start a medical-device company. One brisk morning in early 2008, it all changed.

I was reading articles in the *Financial Times* and other business magazines online. They were analyzing the rickety state of the U.S. economy. Each article offered its own diagnosis and prescription different from all the rest. Those news items corroded my confidence. Why? As a freshly minted MBA, I had no idea what they were talking about. They were unlike anything we discussed in economics or finance class. They defied everything I thought I knew. I felt as if I had reached a state of intellectual renunciation. I needed to *unlearn* in order to relearn the basics.

Later that morning, I went to my lab to do clinical studies. On one of my study subjects, I rigged up sensors to monitor the cardiovascular effects of a noninvasive technology that we were developing. Our research was centered on stimulating the calf muscle pump to improve lower-leg circulation. More than three-fourths of the blood volume in the human body is below the chest. For the blood to flow back to the heart effectively—on every heartbeat, against gravity—the veins need to be compressed by the contractions of skeletal-muscle fibers. That's why the calf muscles are also called the "second heart" and their inadequacy is implicated in many chronic health disorders.

As I was monitoring the peaks and dips, and the drifts and

shifts of the beat-to-beat blood pressure data, they reminded me of the fluctuations in stock prices. My understanding of the physiology of the human system began to converge with my *lack* of understanding of the way the financial system worked. I had an epiphany. Instead of stimulating the calf muscle pump, I thought I should start stimulating the economy.

Some web searches later, I decided to apply for an economic policy fellowship at the National Academy of Sciences. I didn't even tell my PhD adviser. Two of my mentors, inured to my crazy ideas, offered to write recommendation letters. Everything seemed intuitive—until I had a panic attack after I submitted the application. The abruptness of my decision kept me restless. I convinced myself that I had a vanishingly small chance of being selected as a fellow. In the days that followed, life trended back to normal.

A few weeks later, I was short-listed as a finalist for an interview. Soon after, I was selected for the fellowship. In the autumn of 2008—at the height of the global economic crisis and a historic presidential election—I took a semester off from university and went to Washington, DC. It was a turning point in my life. As someone with a business degree and practically no working knowledge of the economy, I was receiving a ground-level introduction to the intricacies and malaise of economic policies at play. I was fortunate to work for an advisory board chaired by an influential economist who was a former treasury secretary.

Debates during an executive board meeting introduced me

to the head-spinning issues of the day. It was as if I had been plunked down at the main control panel of a space shuttle's command module. Topics of discussion included finding the right blend of trade policy, fiscal policy, monetary policy, corporate incentives, federal research support, and several other options needed to keep the economy humming. The experience in Washington made it clear that my education had been disconnected from the fragilities of the real world. It was bewildering.

After all, who was I to judge? I had just floated from the fresh waters of engineering to the salty waters of public policy.

5 ENGINEERS AND ECONOMISTS hail from different disciplines, but both professions are rooted in rationality and quantitative rigor. From new products to new policies, engineering and economics routinely rely on the principles of optimization—again, achieving a desired objective under a set of limitations. Harvard economist Gregory Mankiw has argued that "the subfield of macroeconomics was born not as a science but more as a type of engineering," considering that the originally practical orientation of economics seems to have changed over time.

At least two concepts of optimization overlap between economics and engineering. The first is *utility maximization*. If we go back to the example of congestion pricing, there was hardly anything new about the idea of levying people for using public utilities. If something is in short supply, then you can charge more for it. But in applying the principles of utility maximization, the IBM engineers effectively reduced the traffic congestion mainly through behavioral shifts, with modest change to the existing infrastructure. Perhaps the same reasoning could be applied to Gribeauval, whose goal was to maximize the utility and impact of his modular cannons.

The second concept is *mechanism design*—which is considered "the 'engineering' side of economic theory"—as economist Eric Maskin put it in his 2007 Nobel Prize lecture. How can we design a "preferred" mechanism to achieve a broad social objective? In the case of IBM, traffic reduction was achieved by people's eventual buy-in of the digital tolls. Even more, with Google Maps' traffic prediction technology, people may plan to use a different mode of transportation in advance—which could have significant time and revenue ramifications. Ironically, engineering—the very profession that created the automobiles that produce congestion—was also an "invisible hand" that drove economic transactions and, by way of new technologies, reduced the costs and effects of traffic.

The main difference between the economic way of thinking (which is largely theoretical) and the engineering way of thinking perhaps resides in how ideas are implemented. The economist John Maynard Keynes once stated, "If economists could manage to get themselves thought of as humble, competent people on a level with dentists, that would be splendid." Although Keynes was stressing the need for a practical mind-set among his colleagues (to a certain degree), we might concede that filling a molar cavity is not in the same league as reducing the federal debt or deficit.

In the world of economic policy, several engineers have gone largely unnoticed in the implementation of utility max- imization and mechanism design. Marcel Boiteux, a preemi- nent French engineer, presented a formula for pricing a service when demand is highest. His optimization challenge was to minimize peak-time electricity consumption. Just as roads have capacity constraints during peak hours, so do power plants. Their production capacity is fixed, and the demand could be managed if people consumed less power during peak times. The situation opened up possibilities for an engineer to think like an economist by providing people the necessary incentives to eschew power usage.

"This shift from 'engineer to economist' introduced new ways of reasoning, based on the search for an economic opti- mum," writes Alain Beltran, a historian of the French power industry. This type of thinking started radiating outward as decision makers everywhere quickly realized that peak-hour consumption of valuable resources was ubiquitous. "It's just all over the place from really simple things, like why do some restaurants charge you more for dinner than lunch, even though it's the same meal, probably prepared by the same chef," says Charles Phelps, an economist at the University of Rochester, New York. "It's also about running a parking garage efficiently. You don't want to charge too much for parking on the weekend because the garage is mostly empty anyway."

The airline industry offers a fabulous example of peak-load

pricing. At any given time, airplane capacity is fixed. The carriers can't really add more planes as needed for peak-hour service on Monday mornings. In contrast, they have too many underused airplanes on Saturday nights. Business travelers may have less flexibility and are willing to pay more than other customers. All these factors lead to price elasticity, creating the need for *differential* pricing for the same service.

If you look carefully during your next visit to the supermarket—or your favorite online shop—you'll find that differential pricing is pervasive. Think of razor blades. On their own they are expensive, but when you buy a razor, you may sometimes get two or three free blades as part of the package. Hairdressers charge more for women than for men. Theme parks charge less for a package of rides than for a single ride. Concert halls charge extra for the price of a single ticket relative to the cost of a season ticket. These price discriminations depend on time, convenience, and factors other than the cost of the service itself. When we grudgingly pay the premium to air carriers to get tickets for a summer vacation, it's easy to feel like we're being exploited. We're not. It's a simple rule of optimization to prime our behavior.

Three

ENHANCING EFFICIENCY AND RELIABILITY

(1)   "PIGGLY WIGGLY." For a supermarket the name
was unusual. In 1916, its founder, Clarence Saunders, christened the first shop in Memphis, Tennessee. People
walked past a wooden turnstile, picked up baskets, collected
groceries, and then paid for the goods at the cash register and
left. This was radical. There were no white-aproned grocery
clerks, and the shop had far fewer employees than what was
usual at that time. His shop "cut out all the frills of merchandising," Saunders crowed, and "every forty-eight seconds
a customer leaves Piggly Wiggly with her purchase."

Saunders was a portly man who had worked for a grocer as
a teenager, making four bucks a week. "He liked to preach,"
a journalist observed. "The evangelical strain was strong in
him as in most rural Americans brought up on protracted

meetings, revivals." An autodidact, he became an engineer by practice, not formal training. Frustrated with the inefficiencies of traditional grocers, Saunders studied his own shop's operations from different angles to better understand its labor demands at various times of day. The shop might appear crowded from the front entrance, but looking down from the shop's gallery gave him a precise idea about bottlenecks in the process. In observing the shop from this vantage point, he plausibly carried out a version of what engineers would call *time and motion studies*, which include techniques to improve the efficiency of an operation. The goal is to convert every motion of the worker—and the time associated with that motion—into useful work or revenue. The solutions were in front of Saunders. All he had to do was shift his perspective.

Saunders designed his shops to channel the customers in predetermined ways. All Piggly Wiggly shop had the same color scheme and typeface, and were kept punctiliously clean. Inside the shop, he created a new way of organizing products. Consistent with modular systems thinking, Saunders divided his shop into three distinct areas: lobby, salesroom, and stockroom.

The salesroom was the shop's linchpin. It was subdivided into aisles that featured different products; for example, perishable produce was separated from packaged foods and bathroom supplies. Saunders helped set up the fixtures and lighting system for each aisle. "Every item is plainly marked," read one of Saunders's advertisements. "No clerks to argue with you, trying to persuade you into buying what you don't want to buy. You can wait on yourself in a hurry, or you can be as slow as you desire to be . . . nobody will ask you why you didn't buy anything."

This arrangement significantly improved the shop's efficiency: sales increased up to fourfold, declares a patent issued to Saunders. The shop boasted of an overall stock with four times the variety of ordinary shops, and the prices were very competitive. "In his hand, Piggly Wiggly was never simply a shop but a 'person' who was 'going to be raised on a scientific basis, with a scientific diet for each meal,'" according to Mike Freeman, a biographer of Saunders. By 1923, the Piggly Wiggly chain had expanded to more than twelve hundred shops. Customers were empowered to shop for themselves, but importantly, they had an incentive to do so because lower overhead meant lower product prices. The age of self-service was born.

Piggly Wiggly was a simple but extraordinary retail concept. By the late 1930s, mechanization and automatic-checkout counters had reinforced the growing practice of self-service. Efficiency was the goal. Grocery revenues climbed and distribution costs tumbled. With later developments in point-of-sale systems, bar codes, massive standardized layouts, and gargantuan parking lots, supermarkets came to be known as "category killers." The nature of competition—and collaboration—among customers, shops, and product developers changed forever.

IKEA offers a modern-day take on Saunders's basic concept. The company's philosophy of minimalism, expressed in its modular products, treats all customers like production engineers, letting them assemble their own furniture. "IKEA is Legos for grownups, connecting the furniture of our adulthoods with the toys of our childhood," writes Lauren Collins of the *New Yorker*. The IKEA catalogue "combines the voyeuristic pleasures of browsing albums on Facebook (peeping into other people's houses) with the aspirational ones of *Architectural Digest* (we are all a \$39.99 bookshelf away from being well-read Swedish architects). The IKEA catalogue is a self-help manual for a certain kind of life."

In its subliminal form, self-service connects with our consciousness—an invisible GPS that's navigating and modulating our desires amid neatly lined-up products screaming for attention. This effect, which Piggly Wiggly and IKEA achieved, can be understood from an engineer's angle as a basic output of efficiency. Sam Walton, the founder of Wal-Mart, fully credited the concept of self-service as a powerful reason for his company's success. As we'll discuss in a bit, there was an unlikely beneficiary of this powerful concept.

It was a car company.

(2) JOHN SHEPHERD-BARRON was an old-fashioned Scotsman with a questioning mind. One afternoon in the mid-1960s—as the story goes—Shepherd-Barron was a few minutes late to the bank as it was closing for the weekend. He urgently needed cash. The manager refused to reopen the branch, despite his pleas.

A dyed-in-the-wool engineer, Shepherd-Barron thought he should have the freedom to withdraw cash anywhere and anytime from his bank account. He was the managing director of a currency instruments firm. He worked first on the printing side of the business, and then on the armored transportation side. His next step was to find a way to dispense money automatically. By inventing the automated teller machine (ATM), he completed the circle. How did he do it? "I hit upon the idea of a chocolate bar dispenser, but replacing chocolate with cash," said Shepherd-Barron.

◆   ◆   ◆   ◆

IF NECESSITY IS the mother of invention, then who's the father? The idea of an ATM may seem to have arrived out of thin air, but perhaps not. Some cognitive psychologists use the grandiose term *opportunistic assimilation* to explain the nature of breakthrough insights. Having your mind prepared to exploit an opportunity is an important precursor to spotting one. This intellectual alchemy contains a subconscious association of life lessons and experiences.

For engineers like Shepherd-Barron, what truly comes in handy is the powerful notion of *backward design*—the ability to preimagine the desired outcome and work in reverse to achieve that goal. An epiphany, then, actually results from conscious, methodical planning that supports a confluence of ideas, experiences, and opportunities. Lehigh University's Tom Peters has used the term *matrix thinking*—comparable to the moving around of ideas across the rows, columns, and diagonals of a conceptual matrix—to define an orderly process of spotting, incubating, and combining ideas from various walks of life, and then converting them into practical solutions.

Thomas Edison was incomparable in the ways he capitalized on matrix thinking. The concept, although unknown to him by that name, strongly influenced his attitude toward new opportunities. Technology historian Bernard Carlson has examined Edison's processes as an inventor and notes that his sketches were a "nightmare to study." For example, while developing his version of the telephone, Edison didn't annotate any of his diagrams.

To make meaning out of the mess, Carlson took an approach similar to that of a paleontologist, treating each of Edison's sketches as a fossil. He swept through Edison's portfolio of patents and products, looking for connections and similarities—from superficial mechanical appearances to deeper operational inspirations—hoping to arrive at a "common mental model" to describe Edison's thinking. It was clear to Carlson that Edison was not pursuing a single product, but approaching five lines of inquiry simultaneously. One example is how Edison tried to stimulate current flow by using sound waves to activate an electrical conductor in a magnetic field. Throughout his work, Edison demonstrated adeptness in interchanging ideas and tools, just like the cooperation between the evolutionary processes of variation and selection—a concept that resembles Gribeauval's parameter variation. "These transfers were often like the grafts that plant breeders make, and for Edison they often resulted in improved performance in the telephone under study at any particular moment," Carlson observes.

A key difference in this analogy is that evolution is not goal oriented, whereas engineering is. In that regard, Edison's creations were more of an artificial selection than a natural selection. He exploited the most promising avenues for product development by being "not simply a breeder in the old-fashioned sense but actually more of a genetic engineer," Carlson adds. "Unlike the traditional breeder who must work with the basic biochemical make-up of the plant or animal species, Edison was able to change substantially the make-up of a particular telephone." Over time, Edison created new hybrid technologies, with each version of the telephone performing better than its predecessor. At an even higher level, Edison was branching out like a tree, at times, as demonstrated by dozens of concept sketches with clearly defined goals. "Edison was not simply investigating one kind of telephone," Carlson concludes, "but rather a network of possibilities."

Applying this sort of logic to the development of the ATM, we could say that goal-oriented thinking helped shape a highly focused function: a reliable way of dispensing cash. In imagining and parsing the system and modules of the ATM, from security to data storage, Shepherd-Barron could have worked entirely backward to form a framework for what we now call telematics—a system of systems that unites computing, telecommunication, and transportation technologies.

The first ATM was unveiled in 1967 by Barclays in North London. The four-digit PIN number—a global standard of brevity based on how much information people can effectively remember—was an idea from Shepherd-Barron's wife. Before debit cards came into existence, ATMs processed only checks with carbon-14 radioactive encryption. Public trust in ATMs grew immensely after their reliability was demonstrated time and again around the world.

Another feature of the ATM is that it was not as much a design-based invention as it was a *function-based* invention. If Shepherd-Barron's interest lay purely in the design, his possibilities would have been limitless. The ATM could have taken any shape or form or color. Under the constraints of his final objective—a chocolate bar dispenser that instead dispensed cash—a design-based approach would have been inefficient. A function-based approach would have made it easy to track progress toward the eventual goal. Testing and retesting of ATM functions anchored the core performance needs of reliability, privacy, and security.

Psychologist Gary Bradshaw has written about the importance of function-based design in the development of airplanes. Wilbur and Orville Wright took about four years to implement their first prototype of a flying machine. While their competitors focused on the design of wings, fuselage, and propulsion, the Wright brothers were devoted to getting the fundamental functions of lift, thrust, drag, and yaw correct. Consistent with the notion of modular thinking, they solved each puzzle at a subsystems level before they moved to the next layer of the assembly, and along the way they invented new instruments and measurement techniques.

Consider the following example among the many conceptual obstacles the Wright brothers faced. Most of their contemporaries thought that flight systems operated in a two-dimensional way, "as though an airplane were going to be a cart running on a road or a ship running on the sea," explains Tom Crouch, a senior curator at the Smithsonian National Air and Space Museum and author of *The Bishop's Boys*, a magisterial biography of the Wright brothers. Other developers thought "about the notion of an inherently stable flying machine—a flying machine that would, if it were struck by a gust, return itself to a stable position automatically." The Wright brothers saw it as a completely different challenge. "From the beginning," Crouch adds, "their goal was to devise a control system that would give them absolute command over the motion of a machine in every axis all the time." Control superseded stability. "That's not so surprising—they were cyclists, after all."

The Wright brothers were also befuddled by another challenge. Their propeller performed well in practice, but not in theory, forcing them to discover a concept base. Their biggest breakthrough came, as Crouch points out, "when they stopped to think about the problem and said that essentially a propeller isn't an air screw at all; it's not like a screw going into wood. It's much more like a wing; it's developing lift. Rather than moving forward through the air, it's rotating, and the lift becomes the thrust that moves the airplane forward."

This was the Wright brothers' version of structured visual thinking: imagining the propeller as a *rotary wing*. "One literally has to 'see' the propeller as a wing moving in a spiral course to make this intellectual leap," Crouch notes. Hurdles aside, the Wright brothers' ultimate goal was reliable, flyable functionality.

The functional orientations of Shepherd-Barron and the Wright brothers (or Edison) have one thing in common. It's what the Santa Fe Institute scholar Brian Arthur calls *deep craft*—the ability to know intimately the various functionalities and how to effectively combine them. "It consists in [knowing] what is likely not to work, what methods to use, whom to talk to, what theories to look to, and above all of how to manipulate phenomena that may be freshly discovered and poorly understood," Arthur writes. Systems-engineering approaches that underlie the efficiency and reliability of low-failure-tolerance products like ATMs and airplanes have a strong connection to deep craft.

In contrast to Edison's propensity for documenting and protecting his ideas, Shepherd-Barron's greatest legacy was possibly that he didn't patent his invention. He didn't want to reveal information about the security-coding system protecting the bank accounts that might enable criminals to crack the code. He chose to keep it a trade secret so that the technology could grow unencumbered by patents. "The power of the ATM is in its simplicity that capitalizes on the much older social technology of cash itself that's been around for 27 centuries," says Michael Lee, CEO of the ATM Industry

Association. "That's why roughly every eight minutes, there's a new ATM being installed somewhere in the world." Paul Volcker, former chairman of the U.S. Federal Reserve, said it best: the ATM is the single most important innovation in the financial industry.

3   IN 1956, a small group of Toyota executives visited the United States to tour the operations of the Ford Motor Company. The delegation included Taiichi Ohno, a mechanical engineer. Ohno was on a "go and see" mission in the midst of a Japanese recession after the Second World War, a conflict that had crippled his country's manufacturing sector.

Ohno was amazed by the colossus of Ford's revolutionary assembly line manufacturing process but still found it inefficient. Why? Ford had excess inventory. And because the company produced more than consumer demand dictated, it had to heavily market its products to move them out. General Motors was in the same situation; its approach was also out of sync with customer needs.

Informed by his experiences in the hand-loom business in Japan, Ohno's instincts ran contrary to Ford's. Why overproduce, stock, and wait for customers' orders? Ohno reported back to Eiji Toyoda, then a top executive who would eventually go on to become Toyota's CEO. Eiji was a tough person, devoted to the doctrines of efficiency. With focus, one can squeeze water from a dry towel, he's known to have said. This sentiment harked back to Sakichi Toyoda, the founder of the Toyota group of businesses.

Sakichi was a self-reliant man. He read and reread *Self-Help*, an 1859 book by Scottish reformer Samuel Smiles. As a loom machinist, Sakichi had rigged a weaving machine that would stop if a thread broke. This improvement opened up the possibility for automation. One person could oversee a large number of looms. Sakichi's business prospered.

In the early 1920s, Japan was struggling to expand its economy. A cruel 7.9-magnitude earthquake had devastated the country's Kantō Plain. Tens of thousands of people had died. The railway system was shredded and the rest of the transportation infrastructure was equally damaged. In the face of destruction, the Japanese people embraced self-help and optimism. This mantra inspired their businesses.

Sakichi's son Kiichiro Toyoda took reign of the family business in the 1930s. Kiichiro applied his experiences from hand looms to start a small car company. From this modest beginning, how did Toyota ascend to become a top leader in auto products?

One critical moment over the years was Ohno's visit to Piggly Wiggly when he was in the United States to visit Ford.

◆　◆　◆　◆

SELF-SERVICE WAS just in time. Self-help meant efficiency. At Piggly Wiggly it was a common streamlining practice to restock only after customers had purchased the products.

Beyond the realm of engineering, this folk approach seems to have helped the legendary nineteenth-century French chef Georges Auguste Escoffier, who—as culinary journalist Bee Wilson highlights—"organized the kitchen into separate sections for sauces, meats, pastries." As a result, Escoffier helped dramatically transform restaurant cooking, and also create "a certain philosophy about what food should be."

The tried-and-true, just-in-time logic of Piggly Wiggly inspired Toyota to minimize the overstocking of parts and tools on the production floor. The official Toyota Production System led to prolific triumphs. Subsequent goals of this approach, such as low-defect manufacturing, ruthlessly focused on continuous improvements to production efficiency. This was called *concurrent engineering*.

"In systems thinking it is an axiom that every influence is both *cause* and *effect*," as engineering and management consultant Peter Senge puts it in *The Fifth Discipline*. "Nothing is ever influenced in just one direction." A prerequisite for enhancing efficiency is to recognize the exposed and hidden paths in a process, and their patterns and relationships within a system. Using the metaphor of seawater level, researchers Yuji Yamamoto and Monica Bellgran from Sweden note that in the Toyota model, "when the water level is high, the objects are hidden under the water. By reducing the water level, the objects are brought up to the surface." Concurrent engineering helped expose manufacturing defects, and every challenge had to be addressed with a "sense of urgency."

In recent years, Toyota's approach to waste reduction has

captivated the airline industry. Commercial airlines have adapted their own versions of a "systems approach" to make across-the-board cuts in aircraft weight and fuel consumption. The results include creating cheap, lightweight alternatives, even reducing the size of cutlery to save a few grams on each spoon, knife, and fork.

A trick to improving efficiency, Ohno writes in his book *Toyota Production System*, begins by asking a simple question: *Why?* Repeating that question five times takes you close to the root cause of any particular problem in a process. Here's a sequence of questions, for example, in Ohno's words:

1. *Why did the machine stop?*
   There was an overload and the fuse blew.

2. *Why was there an overload?*
   The bearing was not sufficiently lubricated.

3. *Why was it not lubricated sufficiently?*
   The lubrication pump was not pumping sufficiently.

4. *Why was it not pumping sufficiently?*
   The shaft of the pump was worn and rattling.

5. *Why was the shaft worn out?*
   There was no strainer attached and metal scrap got in.

Senge might see this reasoning process as an effort in understanding "circles of causality" linking to the notion of multiple influences on a system. In a broader sense, thanks to

Toyota, concurrent engineering served as a core transmitter of an idea that grew from a production protocol into a useful management philosophy. Concurrent engineering invigorated various manufacturing sectors, enabled new service protocols, started a work-flow revolution, and opened new pathways for the spread of technologies.

4   ATMS OFFER US a three-way extension: an extension in time by keeping banking open beyond branch hours, an extension in space by taking banking beyond branch locations, and an extension in convenience, enabling cardholders to withdraw cash anytime and anywhere in the world, which was Shepherd-Barron's original vision.

The operating principles of ATMs—let alone automobiles—are based on reliability. ATM failure rates have significantly declined over the past several years, thanks to concurrent error detection algorithms in the software that processes transactions, and to redundancies in the ATM network. The security features of ATMs have improved enormously. Our transactions are secure and ultrafast, even if an ATM has to query a group of connected systems that may not be anywhere near the machine we're using.

Now imagine how a huge financial corporation must be dealing with risks that surround it. "We are gigantic systems operators," says Chad Holliday, chairman of Bank of America and former CEO of DuPont. Information security is extraordinarily crucial for dealing with any form of cyber threat.

These solutions need to be stable and flexible, requiring the highest level of security and several layers of backup. "If these systems were ever to be compromised—say, all the money in people's accounts was wiped, you can imagine the panic. So everything has to be done right," Holliday adds. Whether it's a theme park ride or a bank account, the general principles of safety are the same. As a design requirement, engineers always need to take extra precautions, include failsafe options, factor in backups, and establish redundancies. Indeed, as one joke goes, that's probably why engineers wear both belts and suspenders.

A technical disaster could include a number of malfunctions, but good engineers focus on finding and fixing the root cause. Each failure has a destiny of its own and provides a lesson for future generations. The sinking of the RMS *Titanic* was a grave human tragedy with its root cause in flawed bulkheads that succumbed to the onslaught of the Atlantic Ocean during the ship's maiden voyage in 1912. Further, in the name of silly cosmetics (namely, not wanting to clutter the view from the deck), blended with the arrogant and destructive attitude that the *Titanic* was "unsinkable," the number of lifeboats on board was woefully inadequate. The result was a colossal systems failure.

These were deliberate, if not fatal, design choices—what engineers would call *aggressive trade-offs*—where several other factors overrode safety as a top priority. In contrast, the opposite concept—*conservative trade-offs*—helped tremendously improve the safety features of successive naval systems, and even larger cruise and container ships. Failure is inevitable, but keeping a system as safe as possible is, at best, what any machine or human can do.

For a sports car enthusiast, safety may be expressly traded off a bit with an eye toward performance gains in horsepower. This doesn't mean that aggressive trade-offs are a negative design strategy and conservative trade-offs are the most prudent ones. Several hybrid engineering principles—between aggressive and conservative trade-offs—feed into automobiles designed to meet various customer preferences. Their overarching goal is ideally the same: avoid failure in the best possible way.

In the business of manufacturing spacecraft, for example, "you don't get recalls like in Detroit," says Norman Augustine, retired CEO of the aerospace and defense company Lockheed Martin, referring to the recent large-scale recall of faulty air bags from Japanese manufacturer Takata Corp. "If there's a problem, you can recall six million cars, which is terrible, but in our business, if the damn thing blows up, you don't recall anybody." He was talking about how the aerospace industry lives and breathes the concept of reliability. "There's a certain amount of hand waving with most things in business that you can get away with. With Mother Nature, you can't do that. If you get it wrong, you'll pay for it and you'll pay for it every time," Augustine said matter-of-factly. "As an engineer you'll be judged honestly."

There's a relentless pressure on engineers to get everything right when human lives are at risk. Whether driving on a bridge or relying on a medical device, the last thing we want is a moment of unreliability. "When I used to live in Texas, they had chili contests," Augustine said. "The judges would like your chili and not someone else's. I mean . . . *for goodness' sake!*" he grumbled. "In the aerospace business, maybe everyone in this world likes your rocket but if Mother Nature doesn't, it's all over. You are doomed."

Ensuring reliability is a tricky business, since uncertainties permeate our lives. All an engineer can do is consider all sources of uncertainties and try to minimize them while keeping an eye on last-minute lessons. "You eventually have to push the button to launch your rocket," Augustine says. "You can't wait until every uncertainty is resolved." It's like going to the moon for the first time; you don't want to land on a boulder, but in the 1960s there was no technical solution for knowing where the rocks were. Reliability of a solution is as critical as the solution itself, and along with efficiency, it's a critical ingredient in the social trust of engineering.

◆   ◆   ◆   ◆

MOST ENGINEERING products are in the form of *High Tech* and *High Touch*—as physicist Michio Kaku has described—complementing human desires and needs. With the ATM, Shepherd-Barron effectively engineered aspects of both High Tech and High Touch in one product. High Tech is the incredible global network of ATM systems—or the seamless banking infrastructure that Chad Holliday was alluding to—that enables us to carry out our financial transactions on virtually any ATM. High Touch relates to the satisfaction of withdrawing cash and putting it in your pocket. Insert your card, enter your PIN, take your cash, and walk away. You're done in a matter of seconds. Now *that's* efficiency and reliability.

Four

STANDARDIZING WITH FLEXIBILITY

1 IT'S EASY TO MAKE one of anything.

In 1928, the biologist Alexander Fleming had something strange happen in his lab. One of his petri dishes containing cultures of *Staphylococcus* was contaminated by fungal mold that had destroyed the infectious bacteria. Fleming named the mold penicillin.

He published a paper in the *British Journal of Experimental Pathology* in 1929 highlighting penicillin's potential as an antibiotic. The initial reception was underwhelming. Why? No one knew how to chemically separate penicillin to make it useful in clinical settings. Fleming almost abandoned his research. Over the next decade, the Oxford researchers Ernst Chain and Howard Florey managed to isolate penicillin and report on its therapeutic benefits, but they couldn't crack the code for producing large quantities. Several other research groups had also been trying without much luck.

The challenge grew more urgent during the Second World War, which necessitated an extraordinary amount of penicillin to protect the health of the Allied forces, but large quantities were nowhere to be found. In 1942, for example, drug manufacturer Merck & Co. used almost *half* of the total supply of penicillin in the United States to treat septicemia, a life-threatening infection, in just one patient. Further, each treatment required umpteen doses because penicillin has a very short life in the human body. To conserve the supply, some physicians even reused the penicillin excreted in patients' urine.

Years later, Fleming was giving a lecture. "It was destiny which contaminated my culture plate in 1928," he said, "it was destiny which led Chain and Florey in 1938 to investigate penicillin instead of the many other antibiotics which had then been described and it was destiny that timed their work to come to fruition in war-time when penicillin was most needed."

Fleming used the word *destiny* three times. The first instance is in reference to his serendipitous finding. The second is in connection with Chain and Florey. The third is anonymous. It "timed their work to come to fruition in war-time when penicillin was most needed." This destiny, as it will become evident to you, was arguably even more important than Fleming's chance discovery.

(2) A WOMAN JUMPED out of a seventh-floor window. She was twenty-seven, 5-foot-3, and weighed 120 pounds. She broke through a pine-board roof on her way down, landing on her head at 40 miles per hour and lacerating her scalp. "The victim suffered abrasions over the dorsal portion of the spine and an oblique intra-articular fracture of the sixth cervical vertebra," noted a 1942 report. The woman survived, and recovered in a hospital later that day.

The report's author, Hugh De Haven, was intrigued that the roof had sustained more damage than the woman. He went on to document seven other cases of attempted suicides or unplanned injuries in an effort to understand the physical limits and tolerance of the human body. De Haven's curiosity stemmed from an accident in 1916 when he was twenty-two years old. He had studied engineering at Cornell and Columbia Universities, after which he applied for a position in the U.S. Army Air Corps. Following a rejection, De Haven volunteered for the Canadian Royal Flying Corps as a cadet pilot.

One day during flying practice, De Haven was in a midair collision with another training aircraft. A 500-foot free fall ruptured his liver, gallbladder, and pancreas. His legs were fractured. De Haven wondered how he had survived when the same accident killed the other airman. Why did the same crash result in different injuries? This question laid the groundwork for the field of crash and survivability analysis that underpins the safety features of modern transportation systems.

Over the following years, to help make automobiles crash-proof, De Haven considered the principles of commercial packaging. Boxes and containers are designed to withstand a variety of forces to protect their contents. As a basic principle, De Haven wrote, a "package should not open up and spill its contents and should not collapse under reasonable or expected conditions of force and thereby expose objects inside it to damage." From there, he built on the concept of "interior packaging" that would help prevent damage to the contents "from impact against the inside of the package itself." And to achieve an optimal level of safety, De Haven added, a packaging engineer "would not test a packing case by dropping it [only] a few inches."

Using a modular thought process informed by structure, constraints, and trade-offs, De Haven segmented automobile systems according to their safety elements: container, restraint, energy management, environment, and postcrash factors. The first letter of each of these elements creates the acronym CREEP, which offered a framework for studies on *crashworthiness*. De Haven's literal comparison of passengers in an automobile to "fragile, valuable objects loose inside a container" eventually led him to patent the design for the three-point seat belt—now a standard automobile feature in most countries.

The seat belt design needed to differ from a shoulder harness—which De Haven knew from his experiences was effective for fighter pilots but not for automobile passengers. While the harness had the advantage of securing the upper torso and limiting "extreme forward movement," it was uncomfortable and overly restricting. De Haven's belt could be comfortably adjusted across the lap and shoulder, minimizing potential head injuries during a crash. Seat belts help save tens of thousands of lives every year, substantially reducing deaths and injuries per mile traveled and thus enormously enhancing highway safety.

Let's go back to De Haven's studies for a minute. His subjects were *voluntarily* jumping out of windows and landing on their heads. So you might ask whether his studies were really "scientific," and some people did call him a crackpot. Acceptable science relies on repeatable results. Purists may argue that De Haven's subjects were anomalous and not representative of a "normal" population. Some of them were, in fact, trying to commit suicide but failed. I doubt if De Haven's study protocols would ever be approved by an ethics committee under current laws. This wasn't science in its pure form, but practice leading to evidence.

It's difficult to "fully appreciate the fact that the head weighs as much as a ten-pound sledge hammer and packs the same terrific energy when it strikes a dangerous object at 40–50 mph," De Haven wrote. "If the head hits a solid structure which will not dent or yield at such speeds, the head itself must yield, and crushing injuries of the skull and brain cannot be avoided. But if the head hits a light, ductile surface at such speeds, even a fairly strong metal surface will dent and bend and absorb the energy of the blow, thereby modifying the danger of skull fracture and concussion." In 1946, De Haven went on to demonstrate in a famous experiment that a 1½-inch-thick cushion would save eggs from breaking even when dropped from 150 feet.

De Haven's observations show how trial-and-error engineering preceded organized science in giving rise to a new system of knowledge. He reconfigured not only how we think about public safety, but also how we think about public health writ large. De Haven's work effectively helped change the practice of seat manufacturers, airplane builders, and automobile designers in making seat belts an integral part of the safety systems in their products. Seat belts have been heralded by the U.S. Centers for Disease Control and Prevention as one of the ten greatest public health achievements.

(3) MARGARET HUTCHINSON was born in Texas. Following her father, she became an engineer, graduating from Rice University, Texas. Later, in 1937, she defended her thesis—*The Effect of Solute on the Liquid Film Resistance in Gas Absorption*—and became the first woman to receive a PhD in chemical engineering from the Massachusetts Institute of Technology.

Hutchinson was also a caring wife and mother. "Heating, cooling, washing, drying—these are all household tasks. But when they are to be done on an immense scale, thought and planning must be made—and this is chemical engineering," she told a reporter once. "Fractional distillation, with closely controlled heat for just the correct separation of hydrocarbons in a chemical plant, is quite comparable to baking a cake," she explained. "And making ice cream at home is much like controlled crystallization in industry."

As a precocious engineer, Hutchinson designed a production process for synthetic rubber and worked on a system to distill high-octane fuel for fighter jets. In addition, she led a petrochemical installation in the Persian Gulf. Recognition of these achievements is what roped Hutchinson into the penicillin mass-manufacturing project.

Extracting penicillin from the mold was no child's play. "The mold is as temperamental as an opera singer, the yields are low, the isolation is difficult, the extraction is murder, the purification invites disaster, and the assay is unsatisfactory," a Pfizer executive complained. This was the task assigned to Hutchinson.

Instead of designing and building a reactor vessel for the chemical reactions from scratch—which meant more time, money, and uncertainty—Hutchinson opted for something that was already functional. Some researchers had found that mold from cantaloupe could be an effective source for penicillin, so she started there. Her team then revised a fermentation process that Pfizer was using to produce food additives like citric acid and gluconic acid from sugars, with the help of microbes. Hutchinson swiftly helped convert a run-down Brooklyn ice factory into a production facility. The deep-tank fermentation process produced great quantities of mold by mixing sugar, salt, milk, minerals, and fodder through a chemical separation process that Hutchinson knew very well from the refinery business.

Hutchinson stoked penicillin production to substantially higher velocity. The interbreeding of two disparate entities—fermentation research and petrochemical process engineering—led to high-quantity production of one of the most important antibiotics ever. The process protocols were improved and standardized along the way. Hutchinson collaborated with mycologists, bacteriologists, chemists, and pharmacists to understand the specific needs of the production system and its outputs. Areas beyond the boundaries of one's narrow expertise are what systems engineers call *adjacencies*.

Once the outcomes were stable and reliable, other pharmaceutical companies adapted Hutchinson's approach to mass-produce penicillin under the direction of the U.S. War Production Board. In the first five months of 1943, deep-tank fermentation processes yielded four hundred million units of penicillin. Later that year, in the weeks before the Normandy invasion, the outputs were prodigiously multiplied *five-hundred-fold*. By August 1945, 650 billion units of penicillin were available for military and civil use. After the war, Pfizer and other pharmaceutical companies took the improved fermentation process "far beyond the art of brewing" to produce other trade chemicals and medicinal products.

4 Hugh De Haven's studies on crashworthiness are examples of how a nonstandardized approach can end up creating a standardized system of passive safety. His tests were conducted during a period when accidents were "looked upon by a superstitious populace as being the result of 'bad luck' or acts of God," safety expert Howard Hasbrook wrote. "This reliance on 'luck' apparently stifled any development of safety engineering or design for the protection of human life in accidents." As automobile use began to increase along with the frequency of collisions, De Haven's approaches to help prevent—instead of suffer—the effects of an accident were avant-garde.

The story of seat belts is a case of evolution, which means that, at one point, the technology was suboptimal. Continued improvements were needed before seat belts could be implemented on a national scale. Moreover, the effectiveness of seat belts would have been limited without political and public activism, as well as aggressive regulation of drunk drivers. But the technical improvements themselves needed to develop gradually. A perfect technology is never possible, and aiming for one is unrealistic—a bias sometimes called the *nirvana fallacy*. Often, great engineering designs are foes of reasonable suboptimal designs.

In addition to relying on evolution, technologies like seat belts support systems integration. On their own, the effectiveness of other independently evolving safety technologies, like roads, sirens, and traffic lights, would have been limited. Only by the resourceful convergence of these systems was a safety infrastructure made possible. This process is not unlike the biological process of *recombination*—one of nature's oldest tricks to produce variation from existing systems. The resulting system may produce additional useful standards that were previously unimaginable.

The engineering practice of recombination has led to legions of *compound technologies* that have used or enabled standardized manufacturing platforms for a great variety of applications. Technologies are combined in different ways for different uses by services, agriculture, clothing, construction, mining, and transportation sectors, among others. How and when they're grouped during the growth of these industries may explain why individual sectors peak in productivity at different times. Moreover, these sectors are also now more tightly coupled than ever.

Perhaps the best example of a general-purpose compound technology is the Internet. The Internet is not one thing, but multiple things pulled together as one. It's an unprecedented recombination of systems like processors, storage solutions, algorithms, and communication technologies, to name a few. A search engine can produce millions of entries on a topic instantaneously. How is it practically possible to integrate all the information from around the world and present it digitally within a fraction of a second? Modern Internet content aggregators—or mash-ups—have become extremely dynamic in seamlessly combining information from various sources in various formats. Though impossible years ago, now this gathering of information is done through the integration of multiple systems, each on its own evolutionary path but crucially relying on a uniform set of standards.

The same analysis could be applied to the recombination of sirens and visual warning systems. Two completely different systems were united and continually refined to produce better results than each could have produced on its own. New developments in sirens, including variable tones and frequencies, merged with flashing lights in red, blue, and white, helped provide a genuine sense of emergency. This coalition of sounds and colors permanently changed the course of integrated public warning systems and also served as a true demonstration of how technology is linked to biology in the form of loudness and brightness. What made this recombination possible was not random convergence, but a deliberate system of standards.

◆   ◆   ◆   ◆

THE NOTION OF standards relates to the principle of *interpretive consistency*. We label things of the world and put them in separate buckets. When you hear George Gershwin or Frank Sinatra or the Rolling Stones, you are able to instantly classify their different types of music because each type has attributes typical to a particular genre. We apply structure— from the cuisines of the world to fashion to SAT scores to medical diagnoses—to inform the syntax of our minds. Standards are for products what grammar is for language. People sometimes criticize standards for making life a matter of routine rather than inspiration. Some argue that standards hinder creativity and keep us slaves to the past. But try imagining a world without standards.

From tenderloin beef cuts to the geometric design of highways, standards may diminish variety and authenticity, but they improve efficiency. From street signs to nutrition labels, standards provide a common language of reason. From Internet protocols to MP3 audio formats, standards enable systems to work together. From paper sizes (A4, for example) to George Laurer's Universal Product Code, standards offer the convenience of comparability.

India switched to the metric system in 1956, almost a decade after its independence, having never had uniformity in measurements. One account noted that the country had more than 150 "local systems of measures" with such dramatic differences that the "post office alone required 1.6 million" weights between 1 gram and 20 kilograms. New standards clearly came in handy. Until recently, lack of standardizing was probably the primary reason that mobile phones didn't work reliably across international borders, but the situation has improved a lot. An absence of standards is not a limitation of engineering capabilities, but a reality dictated by business incentives.

Implementation of better standards and tools of interoperability could help improve health care efficiency and reduce wasted expenditures, to take one example. "My pizza parlor is more thoroughly computerized than most of health care," notes medical quality expert Donald Berwick. "To a large extent, health care systems were not designed with any scientific approaches in mind. Too often there are long waits, high levels of waste, frustration for patients and clinicians alike, and unsafe care. A bold effort to design health care scheduling systems, process flows, safety procedures, and even physical space will pay off in better, less expensive, safer experiences for patients and staff alike."

As we have come to rely on standards, it has become easy to engineer ultracomplicated systems. We see the complexity of systems increasing in all sorts of applications—from smart power grids to nuclear reactors to data clouds. A commercial jet has millions of parts, tools, and components produced by different manufacturers, assembled, perfected, and guaranteed to work during the first flight. This flexibility has removed the intimidation factor from building complicated systems. In fact, we've got so good at creating them that it's harder and harder to understand what simplicity means anymore.

(5) "MR. X. HAS a sore throat. He buys some penicillin and gives himself, not enough to kill the streptococci but enough to educate them to resist penicillin. He then infects his wife. Mrs. X gets pneumonia and is treated with penicillin. As the streptococci are now resistant to penicillin the treatment fails. Mrs. X dies. Who is primarily responsible for Mrs. X's death? Why[,] Mr. X whose negligent use of penicillin changed the nature of the microbe," said Alexander

Fleming, speaking in December 1945 at a high-profile gathering in Sweden. "*Moral*: If you use penicillin, use enough."

Fleming's prescription came during his lecture for the Nobel Prize that he shared with Ernst Chain and Howard Florey. In the wake of penicillin's success, Fleming, Chain, and Florey traveled around lecturing and amassing medals and honors. Several of their colleagues were knighted or elected as members of prestigious scientific academies. Margaret Hutchinson, however, was at home taking care of her toddler son. Billy "keeps us so busy we don't have too much time for outside hobbies," she told a local newspaper. In the amazing story of penicillin, Hutchinson and others who made penicillin available to the masses at a critical juncture are not even footnote personalities.

As a society we are good at celebrating the initiators. Why do we overlook the many ingenious adapters like Hutchinson, whose contributions are equally significant, if not more important than the original discoveries themselves? Adaptation is a preeminent form of creation, though it's seldom recognized at the same level. As historian John Rae puts it, "'Adapt, improve, and apply' may have less glamour than original creativity, but the technique of application may in itself be more significantly creative than the original idea or invention."

Rae's core sentiment about adaptation can be extended to the Renaissance era. Johannes Gutenberg invented his printing press by repurposing a wine press for use with olive oil–based ink and block printing. This approach to mass production created a flexible world standard: books. Literacy levels rose and eventually stimulated new social orders.

Gutenberg may or may not have faced process pressures as acute as those that confronted Hutchinson, or Gribeauval in the French army, who deliberately and systematically adapted existing technologies for his version of cannons. Gribeauval's approach to interchangeability resulted in a new technical construct for the army. But that was hardly novel, since clock makers had been actively using the concept of interchangeability for decades. The Toyota Production System refined the operational principles of Piggly Wiggly, John Shepherd-Barron reimagined a chocolate bar dispenser to create the ATM, and Hutchinson applied existing ideas from refineries to produce penicillin. These engineering approaches are not simply imitations, but novel creations guided by strategic inspiration and purpose.

Another metaphor in evolutionary biology that's useful for thinking about creative adaptations is *transduction*. It relates to the process by which genetic elements of one organism are directly transferred into another to generate new features, just as viruses borrow and transfer their properties across hosts. Engineers routinely exploit this design approach. Henry Ford and his top engineer, Harold Willis, didn't invent the automobile; they *transduced* it. "The way to make automobiles is to make one automobile like another, to make them all alike . . . just as one pin is like another when it comes from the pin factory," Sir Harold Evans points out in his book *They Made America*. By integrating lightweight vanadium steel as the chief production material in their existing assembly line process, Ford and Willis helped unleash a new mass-production paradigm.

Thanks to the vintage concept of parameter variation—which Gribeauval, among others, championed—adaptable manufacturing techniques subsequently guided the standardized production of high-quality drugs, vaccines, soft drinks, and food products, thus demonstrating engineering's central role in economic progress. Accidental discoveries like Fleming's often have little to do with the structure, constraints, or trade-offs of the engineering processes that helped win a war while simultaneously creating jobs, protecting health, and maximizing productivity.

Fleming received a statesman's funeral at St. Paul's Cathedral. For all the right reasons, he was hailed in the U.K. as a "national hero." Comparatively, with no fanfare Margaret Hutchinson died on a quiet winter day in her home in Massachusetts. "I actually received little encouragement when I said I thought I could be an effective engineer as well as a woman," Hutchinson had said years before.

"I'd walk out on anybody who tried to talk me out of my ambition."

Five

SOLUTIONS UNDER CONSTRAINTS

(1) VARANASI IS A MICROUNIVERSE of gods, mystics, and mendicants. An ancient town in north India, it's home to more than twenty thousand temples. The center-piece of Varanasi is the Ganges—or "Mother Ganga," as it's fondly called. The Ganges originates in the Himalayas—the "abode of snow"—and rages and meanders for 1,500 miles, supplying water to 40 percent of India's population in an esti-mated 115 cities.

At dusk, priests and devotees gather for the age-old lamp ritual at one of Varanasi's *ghats*—embankments whose stairs descend into the Ganges. Thousands of worshippers watch this spectacle of devotion as priests chant in praise of Mother Ganga.

In an otherwise noisy Varanasi, silence can be found by the

faithful. Hindu scriptures have noted that in this insubstantial material world, the water of the Ganges and dwelling in Varanasi are two of the most substantial things. "In religion all countries are paupers, India is the only millionaire," said Mark Twain, who described Varanasi as "older than history, older than tradition, older even than legend, and looks twice as old as all of them put together."

I first visited Varanasi when I was eight years old. I decided to go back. With delays, it was a thirty-eight-hour train ride in a second-class compartment from Chennai. The whir of the ceiling fan—and an inquisitive civil engineering student in the next seat—gave me company while the train alternated between electric and diesel locomotives as it traversed five of the largest states in India. From water-starved farmlands to lush green mountains, and from boisterous traffic to persistent *chai wallahs*—tea sellers—in fake designer boot-cut jeans, every aspect of the country appeared in a state of flux, except for the train's rhythmic metronome.

Why Varanasi after all these years?

To meet a holy engineer.

(2) 1874. SMOKY RIVER PASS in the Canadian Rockies. It was fifty below zero and the workers' noses, ears, and toes were frozen. Fatigue and frostbite spared none. Even burning the driest wood produced steam, not smoke. For five and a half months, they had lived off of bread, baked beans, and bacon, while their sled dogs chewed on dried salmon.

After camping near a glacier, when the men set off again, the oldest dog "made a feeble effort to rise, gave one spasmodic wag of his tail and rolled over dead," noted a journal entry. "A hole in the snow on the bank was the only grave we could make for him." These pioneers were building a country.

This was Sandford Fleming's team.

Fleming was born in 1827 in the Scottish lowlands of Fife. After a parish education, he apprenticed his way up to become an engineer. He then moved to Canada and got a job with a rail company. Years later, he became the chief surveyor of the Intercolonial and Canadian Pacific Railway system.

British Columbia entered the Canadian confederation in 1871. The lawmakers of the growing nation were eager to establish a coast-to-coast railway system within the decade. But no one had fully surveyed the continental landscape before. That was the task assigned to Fleming, along with his team, under extraordinarily hostile conditions.

Fleming and his crew mapped about a dozen different routes to and from British Columbia through the Yellowhead Pass. Throughout his surveys, Fleming relied on crude geometric calculations based on longitude, as there was no uniform time across the regions.

"There was no 'system.' Like the rail lines, the different times touched or overlapped at 300 points in the country," according to Ian Bartky, a historian of timekeeping. Halifax and Toronto were divided by five time zones, each differing by tens of minutes. One estimate of the number of time zones between New York and San Francisco suggests 144! Even regionally, timekeeping was in disarray. If it was 12:13 in Boston, it was 12:27 in Philadelphia and 12:32 in Buffalo.

In 1832 the United States had about 229 miles of railways. By 1880, the country had increased its rail infrastructure to close to 95,000 miles. To preserve the sanity of the train driver, each rail company began to maintain its own time. Clocks had up to six dials, and train stations displayed the times in various cities. A train going from Baltimore, Maryland, to Scranton, Pennsylvania, in those days might follow Baltimore time, creating the danger of collisions when trains operated on a single track. Today, when anyone from Okinawa can easily coordinate a conference call with someone in Ouagadougou, this older "system" of timekeeping sounds crazy.

But none of these issues came under scrutiny until a disaster happened.

(3) VARANASI TRAFFIC is shambolic. It's a humbling reminder that the British drive on the left side of the road and we (Indians) drive on what's left of the road. "The traffic is not terrible at all," wrote novelist Geoff Dyer. "It is beyond any idea of terribleness. It is beyond any idea of traffic."

I took a rickshaw pedaled by a lean, muscled man in his thirties. My destination was *Sankat Mochan*—the temple of Hanuman, the monkey god and dispeller of troubles— where Veer Bhadra Mishra was *mahant*, or high priest. Sankat

Mochan is a significant religious institution, started by the sixteenth-century saint-poet Tulsidas, who wrote *Ramcharit-manas*—one of Hinduism's most venerated works. Priests enjoy a special status in Varanasi, and devout pilgrims take their rituals seriously. However, Mishra led a paradoxical life. Being a priest was just one part of his identity.

That evening, on the walk to Mishra's home in Tulsi Ghat, dark clouds descended, and a slanting rain began. As I rolled up my Levis to escape Varanasi's slush and dirt, which flowed fast over my feet into the Ganges, my recorder slipped from my shirt pocket and into the river.

I climbed a steep stairway of seventy-odd steps and arrived at Mishra's door drenched. The priest welcomed me with a warm smile. "I thought you wouldn't be able to come," he said. I walked into his hushed and peaceful visitors' room, which some have called the "throne room." Mishra, silver-haired and mustachioed, was in his early seventies. He radiated calm.

My recorder had been liberated, my camera's battery was dead, and my notebook was soaking wet. My technologies had failed.

Mishra laughed.

"Let's just talk."

◆　◆　◆　◆

MISHRA WAS BORN to an orthodox Brahmin family. His father was then the mahant of Sankat Mochan. When his father died suddenly, Mishra was elevated to the position of high priest at age twelve, and his uncle became his guardian. Per tradition, Mishra learned Sanskrit, music, and wrestling.

Mishra also had a bent for science, though he was not adequately prepared for eighth grade. The local school was reluctant to admit him, but the young Mishra convinced the school principal to do so. Mishra excelled. Later in university, he majored in civil and municipal engineering, and fluid mechanics—the study of liquids and gases in motion— became Mishra's passion. He eventually acquired a PhD in that subject and became a professor and department head at the top-ranking Banaras Hindu University—now a branch of the Indian Institutes of Technology.

Mishra's engineering background combined with his priesthood to give him a deep understanding of the causes and possible solutions for the serious pollution of the Ganges. Through a combination of facts, faith, and activism, Mishra began working on cleaning up the river. "Ganga is not the mightiest of the rivers, nor is it the longest," Mishra said, his hands flitting as he spoke. The Ganges supports the livelihood and traditions of one-twelfth of humanity. "Please tell me if there is this kind of relationship with any other river in the world."

4) JULY 1876. Sandford Fleming was traveling in Ireland. He reached the Bundoran station around 5:00 p.m. to catch a train to Londonderry. The ticket said simply "5:35," so Fleming assumed the train would arrive shortly, but eventu-

ally he realized that the train was not to arrive until 5:35 *a.m.* Fleming stayed overnight in the station and missed his ferry connection to England.

A contemporary of Fleming wrote with exasperation, "The traveler's watch was to him but a delusion; clocks at stations staring each other in the face defiant of harmony either with one another or with surrounding local time and all wildly at variance with the traveler's watch, baffled all intelligent interpretation." This confused practice was the norm until a universal twenty-four-hour construct, championed by Fleming, came into existence.

Fleming developed his idea out of modular systems thinking. Time zones were systematically separated into one-hour modules for easy coordination across countries. Every 15 degrees of longitude was equivalent to one hour; thus, by circling the globe in 24 hours, one could cover 360 degrees. The zero-degree prime meridian was later placed at Greenwich. Railway companies began adopting Fleming's idea and putting it into practice in 1883. The establishment of time zones spawned new possibilities for those working in astronomy, meteorology, power production, army, and navy—all of which needed a way to systematize time.

Fleming's idea was an instant hit among policy makers. Some critics called him a communist—the same reaction that greeted ZIP codes. An unexpected, but influential champion of Fleming's idea was the president of the United States. Under Chester Arthur's leadership, the International Meridian Conference was held in Washington, DC, in 1884. In 1885, standard time was implemented worldwide.

Time may well be a "bloodthirsty savage"—in the words of Fleming's biographer, Clark Blaise—but standard time was introduced as an idea and implemented nonviolently. It didn't take a war or even a dollar to install a universally relevant system. Indeed, Fleming's idea was as profound as some of the other creations that mark our lives: 7 days, 12 months, 24 hours, 365 days. Standard time is our "culture's time."

5   SHE WAS WAITING to die. An abandoned widow probably in her late seventies, she had a funereal look and what appeared to be Parkinson's. A diagonal crack fractured the right side of her coke-bottle eyeglasses. I was at a makeshift hospice in a nondescript, dilapidated building in Varanasi.

The concrete floor was dark gray and cracked in several places. Smoke residue coated the walls. I knelt down to ask the woman's name. She harrumphed but didn't speak. I gave her three hundred rupees—about six dollars—so that when the time came, she could purchase wood for her own cremation. She put aside her prayer beads and placed her trembling palms on my head. "May Mother Ganga bless you," she said in Hindi.

For many people, Varanasi is just a place to visit. For some, like this woman, it is their final destination. *Manikarnika*

Ghat—the "great cremation ground"—is the last stop. Many Hindus believe that being cremated here interrupts the cycle of birth and death, offering a fast track to salvation. "Death in Kashi [Varanasi] is not feared," as Diana Eck, a Harvard scholar on divinity and comparative religion has written. "Death in Kashi is death known and faced, transformed, and transcended."

◆   ◆   ◆   ◆

SHIVA IS AN UNDERTAKER at Manikarnika Ghat. He has seen it all—corpses of every type. I met him near the cremation grounds by happenstance. He had a sooty complexion, and his teeth were stained by tobacco. He was probably in his midforties—maybe early fifties—and he walked barefoot. He spoke Hindi with his gravelly voice and managed to squeeze in a word or two of English that he had learned from visitors. He agreed to give me an insider's tour of the operations at Manikarnika Ghat.

"That body draped in white cloth is that of an old man," he said, pointing to the pyre burning about 10 feet from me. "Older women are covered in gold-colored clothes, young women in red," he added. "Dead babies, renunciates, pregnant women, and cobras are dropped directly into Ganga," he said. "Water cremation. Direct salvation."

As I followed him, looking left and right, he described how long it took for the bodies to turn to ash. He gave me a tutorial on how he and his coworkers managed the queue of salvation-ready corpses. The ashes and charred remains of bodies are slipped into the Ganges. "Everything goes into Ganga. We *all* end in Ganga," Shiva said with the comfort and confidence that the rituals over these years have given him. Up to four hundred people are cremated every day, he said—dawn or dusk, rain or shine, in flood and in drought.

Shiva then led me past the stacks of firewood. If the woman I met at the hospice can collect enough money before her death, she might be cremated using mango tree wood— the economy-class journey to salvation. Neem (Indian lilac) is business-class, and sandalwood is first-class cremation. Her other option would be an electric crematorium located half a mile upstream, but who knows what might happen there, with the daily power outages. This was the river she has been living to die for. I recalled poet Dean Young's lines:

This is not the river,
it's an explanation of the river
that replaced the river.

Shiva then took me to the small shrine of an eternal fire. "This fire has been burning for thousands of years," he said, adding the word "nonstop" in English. A bare-chested, middle-aged man with shaved head was stepping out of the shrine holding burning dry grass to kindle his mother's pyre. Shiva took some ash from the fire and smeared it right on my forehead.

"May you get your *Mukti* in Manikarnika Ghat," he said.

(6) LATER THAT AFTERNOON, back in Tulsi Ghat, Veer Bhadra Mishra introduced me to his friend G. D. Agarwal, saying, "He has dedicated his life to the protection of Ganges."

Agarwal, an octogenarian, had earned a PhD in environmental engineering from the University of California, Berkeley, decades earlier, before returning to India. He had gone on to become a professor and department chair of civil and environmental engineering at one of the Indian Institutes of Technology. A few years before, Agarwal, a bachelor, had renounced material possessions and become a hermit. He wore an ocher robe, and prayer beads hung around his neck. Protesting the government's decision to build a massive hydropower project in the Himalayas, Agarwal had made national news in India by embarking on a fast until death.

The silence between Mishra and Agarwal was palpable. With some hesitation, I began talking about what I had experienced at the Manikarnika cremation complex that day. Agarwal agreed that cleaning the Ganges was a big challenge. "It's a very complex system," he said in his earthy voice. "One has to think globally but act locally."

Mishra explained that depositing bodies into the river didn't pollute it that much. According to him, these *non-point sources* are relatively minor contaminants, but you can't ignore them. "It's easy to blame these people," he said, but he agreed that behavior change was essential. Between sixty thousand and seventy-five thousand visitors use the Ganges in Varanasi each day. For those who live there, the Ganges is "the medium of life." As a boy, Mishra contracted polio, as well as typhoid, jaundice, and gastroenteritis—all water-borne diseases—but he maintained that "Ganga is our mother. Ganga is our goddess."

Mishra and Agarwal began to explain *point sources*—those causing 90–95 percent of the river pollution. Throughout the course of the river, several outlets discharge millions of gal-lons of domestic and industrial sewage into the Ganges. It's estimated that at some points in Varanasi, fecal coliform lev-els in the Ganges are as much as three thousand times higher than the acceptable level. The World Health Organization confirms that the Ganges has been a source of cholera epi-demics since antiquity. Between the rituals and the sewage deposits, "neither the society nor the government is serious about change," Agarwal lamented, describing the intrica-cies of achieving behavioral change in a consumerist culture. "It may take fifteen years. It may take twenty years. It may take twenty-five years," to achieve what's needed. But from a technical standpoint the cleaning is "not a difficult problem," Mishra attested.

People do, however, get emotional about or disgusted at the state of the Ganges. They often view cleanup of the river as an urban infrastructure improvement project, pouring lots of money into it without tangible results. In 1986, the Indian government launched the multimillion-dollar Ganga Action Plan. As part of the plan's first phase, sewage treatment plants were built on the banks of the Ganges in Varanasi and a few other cities to tackle point-source pollutants. But frequent power outages rendered these electricity-dependent plants dysfunctional. When there was no power, discharges went directly into the Ganges. This still happens today. Mishra was disappointed at the government's narrow thinking. Along with the staff of his foundation's research lab, Mishra began to document the failure of the government's efforts. Technicians monitored and reported daily water quality, which seemed only to get worse.

Scores of independent reports confirmed that the water in Varanasi is unsuited for drinking. Mishra was even more upset—because the scientific commissions and reports were restating the problem without offering solutions. He began a legal battle with the government and spearheaded a mass movement to correct mission drift within the Ganga Action Plan. The government dismissed Mishra's concerns and launched a second phase of the plan in 1994, without dealing with the failures of the first phase. Frustrated, Mishra began collaborating with Varanasi's municipal corporation to make progress. Unfortunately, that partnership soon succumbed to political pressure and collapsed.

Mishra had long proposed the installation of nonelectric, gravity-fed interceptor sewers between the buildings along the ghats. These sewers, in concept, would collect the waste and convey it to a treatment facility supported by an Advanced Integrated Wastewater Pond System—a technology pioneered by the late civil engineering professor William Oswald at the University of California, Berkeley, who was also Agarwal's PhD adviser. In Oswald's technique, wastewater is passed through a series of interlocking ponds where algae fuel photosynthesis, and oxygen helps break down bacteria and other contaminants. The water is then recirculated for use. For Varanasi this is an appropriate technology, but it has yet to be implemented.

Mishra kept organizing community events. He told less educated villagers who visited his temple not to throw waste into the Ganges. Hearing this from a priest was like a message from the gods, and people listened. The effect could not be achieved by technology alone.

The logician Alan Turing once said, "Science is a differential equation. Religion is a boundary condition." His impression was probably that religion is nothing more than a constraint—or even an impediment—to scientific progress, just as a bridle restricts a freely running horse. But in Mishra's view, science and religion were like the two banks of a river, with both essential for its flow. "My campaign has been like a game of snakes and ladders. When it has gained speed, a snake has swallowed it up," he said. "But one day I'll dodge all the snakes. Mother Ganges will help me to save her."

Mishra's persistence began to garner international attention. He was named in the Global 500 Roll of Honor for Environmental Achievement by the United Nations Environment Programme. The *New Yorker* ran a ten-page profile of Mishra in 1998. Shortly after, he was recognized by *Time* magazine as a "Hero of the Planet." Mishra shared the stage with Bill Clinton during Clinton's presidential visit to Agra, home of the Taj Mahal. Clinton was so impressed with Mishra that during his next stop, in Hyderabad, one of India's tech hubs, he said, "There is much to do to protect our planet and those who share it with us. I talked to an engineer who is doing his best to clean up the Ganges River that he worships as an important part of his faith and his country's history."

Despite these accolades, Mishra's foundation still faces challenges. "We started working when our hair was black, and we are still working for a clean Ganga at Varanasi when our hair has grayed," Mishra said several times. There was something melancholy about him. "I pray that I should be able to be intimately connected with my mother my whole life, and that means that I should be able to go to the river, touch her, and offer my prayers," Mishra once said in a documentary. "This is just the reality of the world in which I live."

As he inscribed a copy of *Ramcharitmanas* to me, I asked Mishra whether he would still recommend that I take a "holy dip" in the Ganges. His eyes brightened. "Most certainly. Nothing will happen to you. You can bathe right here in Tulsi Ghat," he said, adding quickly, "Just don't use soap!"

◆　◆　◆　◆

IT WAS A NEW DAY in Varanasi. I passed kids playing cricket with a tennis ball turned the color of coffee. Launderers were washing and slapping clothes on rocks at the edge of the river. The adjoining temples were reverberating with Vedic chants that reminded me of my grandfather's Vishnu temple in our village. Watching buffaloes, cows, dogs, pigs, and human beings share the Ganges revealed a new dimension. The spirit of cleanliness in Varanasi is not about soaps and sanitizers. It is raw and elemental.

One of Mishra's assistants met me at Tulsi Ghat. "Just watch your step," he told me in Hindi, advising me to walk carefully. "The water is shallow here but gets really deep farther out." Standing chest-deep in the water, I looked at the sun and began to repeat a Sanskrit mantra. I then immersed myself completely and took five dips in that dynamic, dark-green water of honesty.

7　FROM TRAPEZE ARTISTS to thoracic surgeons, constraints affect everyone. Yet one person's constraint is someone else's liberty. People diet and try out beach-body boot camps. Governments sequester their budgets and try to find meaning in frugality. Institutions are bound by their protocols and orthodoxies. Religions prescribe and practice constraints. We even apply constraints to get refined results from a web search engine. Yet the goal of these constraints is to reconsider and reevaluate our position in life.

Aeronautical engineer and former president of India A. P. J. Abdul Kalam likes to tell a story from his first year at university. Kalam and six other students were asked to design a light attack aircraft for a semester project. "I was responsible for [the] aerodynamic and structural design of the project. The other five [members] of my team took up the design of

THINK LIKE AN ENGINEER

propulsion, control, guidance, avionics and instrumentation of the aircraft," Kalam recalls.

The project was due on a Monday morning, and Kalam's team hadn't made much progress until the Friday before. Kalam received a warning from his professor that he would lose his scholarship if he didn't pass. As a student from a disadvantaged family, Kalam couldn't afford to lose his scholarship. "There was no other way out but to finish the task," Kalam said. This pressure was crucial to finishing the work on time and on target.

Decades later, Kalam considers this experience a lesson in systems design, systems integration, and systems management, carried out under a rubric of constraints. "If something is at stake, the human minds get ignited and the working capacity gets enhanced manifold," Kalam added. His reflections parallel the words of eighteenth-century essayist Samuel Johnson, who once remarked, "When a man knows he is to be hanged in a fortnight, it concentrates his mind wonderfully." Deadlines and constraints don't suppress innovation; they direct it. When properly used, they may be a gateway to new possibilities.

◆ ◆ ◆ ◆

THE WORLD OF ENGINEERING is full of constraints. *Negative constraints* are imposed by the physical limits of matter. From hardware engineers to airline chefs, and from tennis players to closet organizers, anyone who has worked within a

110
· · ·

defined space knows what I'm referring to. Even within the sheer constraints of nature, engineers pack in new features and functional capabilities while respecting the physical boundaries of technology.

The opposing concept is *positive constraints*, self-organized scenarios that permit new possibilities without the limits of negative constraints. Kevin Kelly, cofounder of *Wired* magazine, discusses these concepts in his insightful book *What Technology Wants*, adding that these "two dynamics create a push in evolution that gives it a direction." Here Kelly is discussing biological evolution, but the idea holds for engineering design as well.

The negative constraints on Mishra stemmed mainly from the politics of vested interests and human behavior that continue to destroy the health of the Ganges. There are three other constraints: the *physical constraint* of laying an interceptor sewer line that might introduce new congestion in an already ancient infrastructure, the *economic constraint* of financing these projects, and the *psychological constraint* of people hesitant to use recycled wastewater for their religious rituals. Traditions often trump logic.

With Sandford Fleming, the constraints were positive. His solution for standard time emerged while he was doing something different: surveying tough terrains for a possible railway infrastructure. His constraint was a brand-new time architecture whose implementation became possible *because of* politics—thanks to the International Meridian Conference. As a lead architect of the Canadian railway system, Fleming presumably faced a number of negative constraints in his other projects. Time and money are obvious and unavoidable negative constraints in our lives. They tend to make their presence known more forcefully than the positive constraints. But in Fleming's example, hard negative constraints like time and money were treated as relatively softer positive constraints, with the outcome being regularized time that could help people save money.

Consider instead the Olympics. The plethora of negative constraints in such a large-scale systems engineering project outweigh the positive constraints. For Sir John Armitt, chairman of the Olympic Delivery Authority for the 2012 Summer Olympics in London, the challenge was to complete the project on time and preferably under budget while managing thousands of subcontractors. For Armitt, the Olympics was more like a lunar mission than, say, running a car company. "In the former case you have the President saying that in ten years we're going to have a man on the Moon. Likewise, in our situation we were told that in five years the 2012 Olympics are going to take place. That focuses the mind entirely on meeting that date and physically creating what is necessary to create," Armitt declared. Time might have been Armitt's principal negative constraint, but London's physical infrastructure surfaced as another—just as Stockholm's layout challenged IBM's approaches in constrained optimization. The business brand and aesthetics of the Olympics were also well established, restricting flexibility.

Software engineers offer yet another view on constraints, using a special exercise called *constraint programming*. If a programmer can reach an expected solution without following a prescribed recipe or algorithm—akin to a jazz concert leaving room for extemporization—then it's an open, or *declarative*, constraint. If the sequence is boxed and defined with explicit rules for the programming approach—as is often the case with performances of formal classical music—it's a closed, or *imperative*, constraint.

In the same vein, computer engineers practice a concept called *denormalization*, which is to think about constraints in reverse. "It's like starting off with an ideal world," says N. R. Narayana Murthy, founder of the technology and consulting firm Infosys. "You design the system as if there are no constraints. Then step by step, you start introducing your constraints and trade-offs." Murthy explained this concept of backward reasoning from the standpoint of new software development, which often has the inevitable constraints of memory, processing power, and final design requirements. The initial condition may look like the vast expanse of the Great Plains, but with the layers of constraints, the landscape is reduced to the pedestrian realities of Fifth Avenue in Manhattan.

8 JOSEPH WILLIAM BAZALGETTE was a Victorian-era civil engineer. He had no medical background,

but as historian Gordon Cook points out, Bazalgette "arguably did more for the health of Londoners in the mid-nineteenth century, than anyone before or since."

Bazalgette implemented the intercepting sewer system that helped release London from the clutches of a cholera epidemic in the 1850s. His design of the pumping station and sewage treatment systems prevented the future contamination of the River Thames. Thanks to him, the Victoria Embankment is now a tourist destination, and not an exposed gutter.

Sewers have become an integral part of the planning and development of the industrialized world, and they are among the most significant and understated public health technologies. Bazalgette's engineering mind-set helped him overcome the challenges of bureaucratic pressures, financial constraints, and a potentially disastrous health epidemic. Ultimately, Bazalgette's ideas resulted in a broadly expandable new public infrastructure that had never existed before.

The Thames and the Ganges can be considered public health hazards at different points in time. To compare the works of Bazalgette and Mishra, let's explore some of their common negative constraints. Bazalgette's deliverables were mainly structural. He was very hands-on and methodical. As the head of the Metropolitan Board of Works, he had a politically sanctioned goal. He led the installation of sewer treatment pumps and the expansion of subterranean networks.

Mishra's challenge required a change in public attitude, arguably a supremely difficult thing to engineer. Mishra's challenge was also to promote a serious form of ecological awareness in a place held together by faith and religion. Though Bazalgette and Mishra approached from different angles, they both strived to create a system of infrastructure that was needed to improve public health.

The worlds of engineering and public health are deeply intertwined. In fact, "the core of public health emerged from engineering," says Harvey Fineberg, past president of the Institute of Medicine of the National Academy of Sciences. "Engineering is the action arm of public health." Of course, clinical medicine and public health are very different in their objectives.

With public health, the concern is the health of the entire population. It's about being *proactive*—preventing something before it manifests. Individual medical care is *reactive*. If public health is about society at large and medicine is about the individual, do they add up to the same outcome? Yes and no. Let's consider first the origins of public health.

Public health has a mixed ancestry. It has deep roots in engineering—sanitary engineering, to be specific—but over time the profession has distanced itself from engineering and moved more toward the medical sciences, especially after Louis Pasteur and others proposed the germ theory of disease at the end of the nineteenth century. If engineering led the "design" aspects of the solutions, then the microbial revolution brought to the fore the "causes" of disease, eventually establishing a scientific knowledge base for the practice of public health. In recent years public health has allowed itself to be influenced by sociology and humanities in an effort to promote community wellness, but this approach may seem nebulous from an engineering point of view.

The constraints in public health are exceedingly complex, since a majority of preventable diseases are a function of *human behavior*—the very factor that vexed Mishra for decades in Varanasi. One reason for this complexity might be, as with engineering, the very invisibility of public health. There's a "lack of drama" in public health, as Fineberg likes to point out. "There are television shows about emergency departments, but will there ever be a show about prevention? Think about the plot line: *nothing happens*."

In our society, drama garners attention. Bazalgette had the drama of the Great Stink of 1858, when the foul smell from the Thames began to upset the lords in Westminster, but for Mishra it was the centuries-old drama of steadfast beliefs and practices. Creations engineered for public health—like curved roads, pavements, dynamic road signs, shatterproof windshields, reduced braking distance, speed displays, air bags, brake lights, radial tires, heat exchangers, and seat belts— were all shaped by constraints. Whether they came about with or without drama is a different question, but they all resulted in dramatic changes to public health and human behavior.

(9) VEER BHADRA MISHRA passed away in March 2013. I returned to Varanasi to meet with his eldest son, Vishwambar Nath Mishra, the current high priest of Sankat Mochan. Once more it rained, and I was drenched as

I entered the visitors' room at Tulsi Ghat. I told Vishwambar about my conversations with his father. We then began to talk about the cleaning of the Ganges. "Nothing is impossible," Vishwambar said. With faith everything can be accomplished. "It's a matter of time. That's all."

Vishwambar's optimism is also fueled by his qualifications as an engineer, which he said he's proud to have inherited from his father. An engineer "knows how to transform and how to introduce new things in a tradition," said Vishwambar, who is also a professor of electronics at the Banaras Hindu University. "I represent my father. My son will represent me. This is how traditions keep on moving," he added. "Basically, we are in a relay race, aren't we?"

Six

CROSSING OVER
AND ADAPTING

(1) AT ABOUT 10:30 A.M. on Saturday, November 13, 1993, eighteen-year old Jennifer Koon finished her shift at a psychology clinic in the suburbs of Rochester, New York. She was excited about that evening: an outing with a friend to hear Billy Joel in concert in Syracuse.

Jennie was 5-foot-4 with blonde hair, brown eyes, and a wide smile. She loved going to pumpkin patches. She had adopted two wolves: Teddy Bear and Cris. Jennie was studying to become a child psychologist at St. John Fisher College, while working part-time at the clinic as a receptionist.

After she left the clinic that morning, Jennie drove her Mazda compact to the Pittsford Plaza. First stop: an ATM machine. With some cash in hand, her next stop was a bakery right across the street. She purchased a dozen assorted bagels to take home.

When she was walking back to her car in the parking lot of the shopping mall, she realized she was being followed by a tall man in his late twenties. A minute later, he grabbed Jennie and threw her in the back seat of her car. Jennie screamed.

She was driven away from the mall and held captive for the next two hours. She was beaten and raped.

Jennie somehow managed to call the police from her wireless phone. But the emergency dispatchers could not determine the phone's location. Jennie's phone did not contain a localization technology—what we now call the global positioning system, or GPS; it would not be available for public use until 1995. Tracking her location was virtually impossible.

Her tormentor kept prodding his half-brother—who had been picked up on the way—"Why don'tcha just do her?"

Then came a gunshot.

The bullet ruptured one of Jennie's lungs, came out through her arm, hit the glass window, and landed between her feet. The half-brother then ran away.

Jennie knew this was her end. The operator kept asking "Hello? Hello? Can I help you?" The entire call was being recorded. At one point, the abductor was seen stopping the car and then banging Jennie's head against the passenger side window.

"Please, please, take me to the hospital," Jennie said.

Another bullet followed—this time to her head.

Later that afternoon, Jennie's body was found in her car in Orpheum Alley, a run-down part of Northeast Rochester.

◆ ◆ ◆ ◆

THE NEWS DRAINED every ounce of life out of Jennie's parents, David and Suzanne Koon. David Koon worked as an industrial engineer at Bausch & Lomb, a manufacturer of vision care products. That Jennie was abducted from a busy shopping mall, raped, and killed in broad daylight with no eyewitnesses was mind-boggling to Koon. He couldn't comprehend it. Neither could the police.

Two weeks later, Koon started his own investigation into Jennie's murder. He visited the mall every Saturday for eight weeks. Like an urban anthropologist, he observed and talked to people. Koon took copious, detailed notes and mapped out the perimeter of the mall. He pinpointed the locations of cameras, magnetic alarms, and even plainclothes security staff in the department stores. To his surprise, Koon found that individual shops had tighter security than the plaza; in fact, one particular shop had more security measures in place than there were in the entire plaza. "That's nuts," Koon thought. "The stores protected their merchandise but not the customers."

Koon approached his county legislator, making a clear-cut case for installing security cameras and closed-circuit television in the mall. No response. Koon thought he could do a better job than the legislator; at least he could respond to letters. So he campaigned and ran for the county seat while working full-time. He lost by six hundred votes—by a narrow 51-to-49 margin for a seat that had been uncontested for fourteen years.

In the months that followed, Jennie's assailant was captured. Eventually, he was sentenced to thirty-seven and a half years to life in prison. Koon was relieved, but he felt his job was not complete. In fact, it was just getting started. He ran for an assembly seat that opened up in a special election. He was out every evening during the harsh winter of 1996, knocking on doors and explaining his motivation.

This time, Koon succeeded.

2   KOON GREW UP in Ellamore, a small sawmill community near the Monongahela National Forest of West Virginia. He lived a hundred yards from the three-room school that covered all grades from first through ninth. The house was heated by a coal-fired potbelly stove. Koon's grandmother fixed lunch for everybody. His dad was a mechanic.

Koon worked in his dad's shop on the weekends, "I always got the job of replacing the exhaust system on automobiles. I was crawling around under the car and the rust would fall in my eyes. I hated it," Koon said. Koon is now in his midsixties with platinum hair; his brown eyes are piercing yet kind.

On a frosty Thanksgiving weekend with low-hanging clouds, Koon and I met for lunch near his home in Fairport at a TGI Fridays alongside the New York State Thruway. Tuning out the high-tempo Lady Gaga remix in the background,

I asked Koon to tell me about his first day in politics as an assemblyman.

This is how it went.

Koon and his wife had breakfast with the speaker of the assembly. "We basically had a nice conversation, but nothing about what was going to happen in session that day," Koon remembered.

The legislative session started that afternoon. Koon was introduced and got a standing ovation. "This is pretty cool," he said to himself. Following the introductions, the members were asked to start voting on bills. "I didn't even know what a legislative calendar looked like, let alone what bills we were voting on," Koon admitted to me. He sat there with no background or briefing materials about the bills. "The lady behind me said, 'Pssst Dave . . . today, just vote the way [that other person] votes because you're a marginal and she's a marginal." So, Koon sat there and watched the board to see how the other person voted, and that's how he voted.

"I didn't even know what a 'marginal' meant. I found out after I asked someone about it; I was a Democrat in a Republican district. That was my day one in politics," Koon said. "That's how naïve I was when I went from the world of engineering into the world of politics."

He had political power without an instruction manual.

"I had no clue."

③ HENRY BADILLO, Charles Wertenbaker, Andrew Melnikov, and Max Guarino were New York teenagers who wanted to form a band. Around 9:30 p.m. on Friday, January 24, 2003, after spending nine dollars on sweets, cookies, and a Starbucks Frappuccino at a City Island gas station, they did something unusual. With their guitars, they hopped into an 8-foot fiberglass dinghy and started rowing into the frigid waters of the Long Island Sound. The air temperature was 33 degrees Fahrenheit.

About twenty minutes later the dinghy started taking on water. One of the boys dialed 911 from his mobile phone. His voice was riddled with fear and confusion. The call lasted twelve seconds. It ended as the boat, and the boys, went down in the water.

The emergency operator at the call center—known as a public-safety answering point—was not able to register the precise location of the accident and decided that there wasn't enough information to notify rescue authorities. There was no way to pinpoint the boys' whereabouts. An appalling fourteen-hour delay in the start of the search operations fanned the flames of public outrage.

The boys' bodies would surface and wash up on shore later that spring.

This was New York City's Jennie Koon moment.

◆   ◆   ◆   ◆

IN ALBANY, David Koon had been championing the need for an improved 911 system for public safety. He knew the nuts and bolts of the GPS technology, but that alone was not sufficient. He needed to *adapt* and play the political game. Learning how to read legislation and how to get a bill passed was like learning a new language and customs of a foreign culture. Several deal-making sessions and draft legislations later, Koon's ideas had gone nowhere. He had redoubtable opposition. The governor vetoed his bills three times.

In the world of policy making, it sometimes takes a tragedy to inspire common sense. Unfortunate as it is, the deaths of the four boys in the rowboat accident served as a poignant impetus for a national conversation on public safety. "I put their deaths on our governor's shoulders," a more politically conversant Koon thundered to the media. Crucially, this boating accident brought Koon's proposed legislation for combining GPS and 911 back to life.

The discussions and arguments in the state assembly and senate reached a political crescendo. Not long after, New York's legislature overrode the governor's fourth veto and passed an "enhanced 911 law." Other states followed suit. New York became the first large state to meet the enhanced 911 standards of the Federal Communications Commission.

Weeks after the law passed, Koon was agitated. No one had a clue how to implement the legislation. "Nobody ever came back to me and talked to me about how to do it," Koon recalled. "They just saw it as a political issue that they could gain a lot of popularity over. They started using *my* cause for *their* reelection campaigns and so on." Much to his dismay, Koon was denied a place on a bipartisan state commission that was to begin implementing the enhanced 911 law. "I was basically shut out because I didn't have enough seniority. I was the young punk on the block that was pushing everything," Koon said. "It really upset me."

Koon started to focus on the technology side of tracking—a subject that his mind had an affinity for. GPS offered pretty high-quality information, and the public-safety answering points would be able to track callers to within a couple of meters of where they were. But the resistance came from phone carriers. They wanted to stick with a triangulation approach, wherein signals are sent to three different towers to pinpoint a person within roughly 200 yards. This level of precision might seem sufficient if you're stranded in a cornfield in upstate New York, but it would be a big problem if you were stuck in Midtown Manhattan. In a thicket of high-rise buildings, the rescue authorities might not even know which building you were in, let alone the exact floor and room.

Koon's fellow legislators seemed to be from a different world. "They had no idea what global positioning systems were for," Koon said. "I mean, I had to teach them starting from latitudes and longitudes!" At the same time, Koon continued to educate himself on the technical aspects of modern communication systems. His vision was foreordained: a robust safety system. He then worked backward—using the engineering traits of structure, constraints, and trade-offs—to achieve a goal often hampered by political hurdles.

Even before the bill was passed, Koon met with legislators individually, trying to move his case forward. He would pose questions like, "If you call 911 now, can the dispatchers tell where you are?" He got blank stares or naïve guesses. "They will not have a clue where you are, because you're on a wireless phone. They know the address of your land line, but your cell phone isn't that way," Koon explained to them.

"And oh my God, sometimes it was like talking to a wall! They didn't understand that this needed to change, and that we had to do something to put this safety system into place," Koon said. "At times, it was very, very frustrating because I was talking to legislators that have been there for twenty, twenty-five years and they think they know it all."

After the law was passed and implementation ramped up, Koon soon faced a barrage of requests for interviews, keynote addresses at conferences, and testimonies in Washington. One technology bulletin profiled Koon this way: "The word 'hero' is too often thrown around too lightly. But if you define a hero as someone who battles the odds, fights powerful opponents and works his way through daunting obstacles to accomplish something that needs doing and to make a difference—well, then, David Koon is a hero."

Over time, people came up to Koon and began thanking him for his leadership. A public-safety answering point official told Koon about an elderly couple who had accidentally driven off into a wooded area. Nobody could see them. When they called 911, the emergency responders located them and dispatched help within minutes.

In another instance, a man fell off a snowmobile in remote wilderness upstate. He broke his back and was lying on an icy slope. Nobody knew where he was. He himself didn't know where he was. Nor had he told anyone he was going out. He was in tremendous pain, and the only thing he had with him was his mobile phone. He called 911. Within the next fifteen minutes he was located and rescued by emergency responders.

"I would have frozen to death out there," the rescued man would later tell Koon.

"You saved my life."

4   IF SEDATING A POLICY issue is easier than solving it, then how does efficiency fit in? As Koon found out, the ideals of efficiency that work well in the tenets of engineering may become spineless in the soap opera of politics. That's perhaps because politics is the art of compromise, and engineering is the art of trade-offs.

"There is nothing more inefficient than the concept of government. It was designed to be inefficient," Koon said. "You cannot remove or redesign the elements you want that are not working because there are vested interests in every sphere of the government. Everyone has a way to justify their own existence." Even worse, in the chronic "obstruct-and-resist" style of operation—what political scientist Steven Smith calls the "senate syndrome"—a disenchanting obsession with "procedural warfare" on legislative proposals quells productivity.

As an industrial engineer, Koon tried his hand at making his own office efficient. He did time and motion studies similar to the ones that helped Clarence Saunders's Piggly Wiggly and the Toyota company's production systems. How much time did each step toward resolving a constituent's inquiry take?

Driven by the data, Koon reorganized his office. He kept a log of every phone call, e-mail message, and letter. He tracked the status and outcome of those requests. The more he learned about the inefficiencies of government, the more he wanted to tackle them.

"I got to know all of the agencies and the people in charge," Koon said. "I then wanted to change the agencies because I had so many constituents coming to me and saying, 'I called the so-and-so department and was on the phone for almost an hour. I talked to ten different people and nobody could answer my question.'"

It took a tenacious decade, lots of patience with the attitudes, platitudes, and bedraggled customs of the legislature, and a host of rejections for Koon to get his bill passed in Albany. "An engineer is a change agent, and here is a system that doesn't want to change," Koon observed.

"It drives me crazy."

✦ ✦ ✦ ✦

IN THE WORLD of public policy, rhetorical bravado is often strangely more powerful than matters of substance. Culture shock aside, many engineers are self-declared introverts. As the old joke goes, an extroverted engineer is someone who looks at the other person's shoes while talking, whereas an introvert looks at his own.

Paul Tonko is an unusual introvert. "I've been looking at my own shoes quite a bit lately, and they've been pathetic," he joked in his office on Capitol Hill. Tonko is one of the very few engineers elected to the U.S. House of Representatives. In the past, as an assemblyman in upstate New York, he partnered with David Koon on the enhanced 911 legislation in Albany.

After coming to Washington as a congressman, Tonko found himself surrounded by lawyers. He'd sometimes joke with them: "You're taught to defend the innocent or the guilty, the good or the bad . . . As engineers, we're taught to decipher what works for this problem, and what solutions are acceptable." In the world of engineering, prudence supersedes popularity. "I'd love to be popular," Tonko admits, "but before I'm popular I need to be right."

Lawyers far outnumber any other professional group in the democratic political systems of many countries. This selection bias depends, of course, on the political culture of a country. For example, in China the phenomena of large numbers of engineers becoming politicians "goes hand in hand with a certain way of thinking," noted an article in the *Economist*. "An engineer's job, at least in theory, is to ensure things work, that the bridge stays up or the dam holds."

Perhaps this is where "politicians and engineers are completely different," says Claire Curtis-Thomas, who migrated from the world of mechanical engineering to become a member of the British Parliament. "We are not of the same world. I think politicians understand us less than we understand them. They probably say we are silly-nilly retentive and don't understand anything." On the contrary, when everyone is narrowing in on deal making or trying to win an election, an engineer is likely to look at the big picture and try to find new synergies. Only certain cultures seem to recognize and encourage that talent.

An engineer's mind is an inference engine. Engineers take clues from nature and apply them to their work. "We fit well within a natural context. We don't seek to argue with things," Curtis-Thomas observed. It was a surprising insight from a politician—a role that inherently relies on argumentation.

Harmony in engineering may well be quite different from political harmony. Engineers are "absolutely wedded—hands, feet, and fingers—to the concept of balance in all things," Curtis-Thomas said. "Everything on the left must balance with everything on the right or it doesn't work; the universe is out of kilter." To truly appreciate the concept of balance, one must be comfortable knowing why something is the way it is, and whether or not it might change its position. In a way, thinking about balance could mean operating like an adjustable stereo equalizer: balancing the volume, adjusting the frequencies, and filtering out the noise.

Just the idea that we can logically break down a system to its essentials while not losing sight of the big picture—modular systems thinking—was almost revolutionary among Curtis-Thomas's colleagues in Parliament. "By and large, [engineers are] seen to be rather dull but hugely reliable," Curtis-Thomas added. "Given the choice between dull and reliable over creative and mostly unreliable, you pick." It's a trade-off.

It may all come down to a form of "inner confidence," as former U.S. Senator Ted Kaufman describes it. Kaufman has degrees in engineering and business management. Before being elected to the senate in 2009, Kaufman served as the chief of staff for Senator Joe Biden. "You're not intimidated by complex problems involving the use of jargon and esoteric words of a particular period that you're involved in," he said. This confidence is intertwined with real nervousness about being wrong in a public setting, let alone in influential lawmaking sessions.

Some journalists and more than a few politicians show no reluctance to make bold pronouncements beyond their areas of competence. Many people with a technical background are uncomfortable with that approach. Why? "In a sense they're cautious," notes Ralph Cicerone, a climate scientist with a background in electrical engineering, and the president of the U.S. National Academy of Sciences. "So, venturing off into politics is not really aligned with the way scientists and engineers think, because they really don't want to make mistakes. The other thing is, they don't want to oversimplify. In the world of governance and politics, even to be able to talk to people, you have to speak in simple language." While that may explain why scientifically trained people hesitate to venture beyond their spheres of competence and comfort, it also underscores the need to blend experience with social intelligence.

"The soft stuff is actually the hard stuff," says Gordon England, who has twice served as the U.S. secretary of the navy, and once as deputy secretary of defense. It's not surprising that many decision makers actually don't like to make decisions, England points out. It's easy to defer or outsource them rather than taking the risk of making a poor decision.

But the challenge in politics is that people are focused on different things. They use different words to describe what they want, resulting in a cognitive bias called *frame mismatch*. Imagine this scenario: You get into an elevator and push the lobby button. The elevator stops and you get out, thinking it's the lobby, but it turns out to be the fourth floor. But you know you didn't push the fourth-floor button. In your mind, when the elevator stopped it was at the lobby. What you didn't know was that someone on the fourth floor had pushed the button to hail the elevator on its way down. That's not part of your mental model, because you can't know that. It's a hidden variable, and politics has loads of hidden variables.

In making the case for the enhanced 911 system, Koon's challenge was to be a three-way communicator. He had to educate his fellow legislators about the need for an intelligent tracking system. He was talking with various technology firms to better understand the communication systems in operation. He was also connecting with his constituents to make the case for their safety and well-being; after all, Koon's reelection depended on their votes.

5 LISTENING IS A fine art. "We go in and try to fix peoples' lives as an engineer. We know how a manufacturing system works, but we have no idea what's going on in their personal lives," Koon said. As an elected official, people came to him seeking help on personal matters—anywhere from having a speeding ticket nullified to changing a specific clause of the divorce law to asking help for a child with disabilities. "I mean, it's just that sort of thing that your ears aren't used to."

Koon is an exemplary listener. With age, he has also developed hearing loss that forces him to pay extra attention. But he honed his listening skills years before he first ran for the assembly seat, when he spearheaded a different community effort. He called it the Rochester Challenge Against Violence.

There were seventy homicides in the Rochester area in 1993. Jennie was the sixty-ninth victim. The Rochester Challenge was unlike anything Koon had done in his life. For others it was community organizing, but for Koon it was an exercise in root-cause analysis. He led hundreds of volunteers in programs to identify sources of violence, to help predict and derail potentially violent behavior.

Taking time off from his day job, Koon came across scores of people marred by conflict. "A woman opened her purse and gave us a butcher knife. She was going to kill her husband after he was asleep," Koon said. He also told me about a ten-year-old boy who was afraid he was going to die. It turned out that the kid was trying to sell snub-nosed .22 revolvers for $25 a pop, and 9-millimeter guns for $110 each on the street. "We got him out of his home. His mom was a drug addict and he was the one supporting her habit."

For Koon, these experiences were revelatory.

◆　◆　◆　◆

CAN THE PRINCIPLES of engineering be applied to politics? It depends. Politics runs on feelings, not formulas. Importantly, as with comedy, timing matters a great deal in politics. As Abraham Lincoln argued in the Ottawa debate of 1858 against Stephen Douglas, "With public sentiment, nothing can fail; without it nothing can succeed."

A good number of engineers tend to put more weight on useful end results than on social interactions—although every engineer would agree that both are important. A good framework to help explain this point is the empathizing-systemizing—or E-S—theory developed by the psychologist Simon Baron-Cohen.

Empathizing engages diffuse neural processes that help us

relate to others. This is in stark contrast to the deliberative, rule-driven systemizing processes that cultivate discipline. A healthy balance between empathizing and systemizing can yield powerful outcomes.

Sometimes the systemizing ability is in overdrive, Baron-Cohen explains. And in reality, such an imbalance may mean a range of social disorders on the autism spectrum. Syndicated cartoonist Scott Adams has succeeded in personifying socially awkward behavior through Dilbert, a bespectacled character who is "an electrical engineer in a big company. He is tremendously skilled and extremely focused in technical things, has almost superhuman capabilities in areas he's interested in, but is weak in some of the more mundane things, like fashion, social convention, manners, and dating."

"Engineering is the paradigm case of such an occupation," Baron-Cohen and his colleagues wrote in a controversial paper published in the scientific journal *Autism*. "This is because it primarily involves a good understanding of objects rather than people." Apple cofounder Steve Wozniak describes it this way: "It's really more a characteristic where you don't socialize. You don't talk the normal language. You kind of feel embarrassed. You're an outsider. You become very scared to open your mouth around normal people. You hear people coming up, doing their talk about 'Hi, nice day,' and the small talk starts up, and you don't even know the clues of how to do it. I don't to this day."

And in the even nerdier parlance of the late astronaut Neil Armstrong: "I am, and ever will be, a white-socks, pocket-protector, nerdy engineer—born under the second law of thermodynamics, steeped in the steam tables, in love with free-body diagrams, transformed by Laplace, and propelled by compressible flow."

◆ ◆ ◆ ◆

PERHAPS KOON'S role model for balancing empathy and rationality was Jennie herself. As he once put it to a journalist from *Newsday*, "Even her voice on that [911] tape was not of fear or hysteria. It was a calm, still-thinking-rationally voice, even knowing what was about to happen. I listen to that tape over and over; that's the bit Jennie taught me about death and dying." Koon still carries that tape in his suitcase wherever he goes. It keeps him grounded, he told me.

One Sunday afternoon some years ago, Koon was driving home from church on the interstate, deep in thought. He heard a mellow voice that he hadn't heard in a while.

"Dad, I'm proud of you."

He immediately pulled the car over to the side of the road. He looked around. He checked the back seat. He rubbed his eyes.

There was no one.

"I lost it," Koon said. "I sat there and cried for fifteen minutes."

Seven

PROTOTYPING

1 STEVE SASSON is a brisk, middle-aged man. At 6-foot-3 he maintains the poise of a cavalry officer. He grew up in the southwest of Brooklyn. His father was a dockworker, and his mother managed their three boys while working as an office secretary. The Sasson brothers are now an engineering triple threat: John Sasson studied civil engineering, Steve Sasson trained in electrical engineering, and Richard Sasson specialized in chemical engineering.

As a kid, Steve Sasson was a natural at tinkering. He built radios and intercom systems in his basement and put up antennas on the roof. He dragged home discarded television sets off the street and took capacitors, resistors, transformers, and tubes out of them.

One Saturday in the mid-1960s, a teenaged Sasson and a

buddy were playing with sulfur, carbon, and potassium nitrate in their basement. Their goal was to make gunpowder. "If it were today, the Department of Homeland Security would be paying me a visit," Sasson joked.

Sasson went to university at the Rensselaer Polytechnic Institute. His first-year physics professor was a quiet, whimsical man, and a renowned educator. "OK, what problems have you had this week?" he'd ask the class. Sasson always had a question, since he struggled with his homework. The professor would go to the chalkboard and "start off with an equation like $F = ma$, and three lines later he'd have the whole thing," Sasson recalled. "It was so elegant. It was so freaking elegant. Even if I had the right answer it took me three pages, but it took him just two or three lines. It was like watching Michael Jordan play basketball. It looked so smooth and freaking easy, but when you tried to do it you couldn't do it."

◆　◆　◆　◆

A FEW YEARS LATER, Sasson read a biography of George Eastman. Eastman, who had dropped out of school, was self-taught. An accountant by trade and an avid kitchen sink experimenter, Eastman eventually revolutionized film photography and founded Kodak. Sasson's core belief about engineering was influenced by Eastman's motto: an artistic technology like the camera should be as "convenient as the pencil."

Sasson joined Kodak.

As one of his first projects, Sasson was asked by his supervisor to explore the potential uses of a new technology called the charge-coupled device. The CCD is an electronic light sensor that was pioneered by Bell Labs. "It was at best a forty-five-second conversation with my supervisor," Sasson recalled. There were no formal reviews or expectations for the project.

Kodak then had a profusion of mechanical engineers. As one of the very few electrical engineers on staff, Sasson thought he should build an image-capturing system "with absolutely no moving parts." As an early-stage technology, the CCD was "very, very difficult to work with," Sasson remembered. Its resolution was 10,000 pixels (or 0.01 megapixel). "On top of the actual device was a folded piece of paper on which the twelve voltage designations were printed," he added. "Next to each one, handwritten in pencil, were specific voltage settings for each pin. At the bottom, it said, 'Good luck!'"

Sasson spent long hours in a back lab doing tests, incrementally inching toward a groundbreaking technology. He barely spoke with his supervisor. "Our plan was unrealistic. No one was paying attention. We had no money. Nobody knew where we were working," Sasson explained. "In summary, the situation was just about perfect!" A year later—in 1976—the twenty-five-year-old Sasson finished a prototype. It was a clunky contraption, more like an 8-pound toaster, requiring sixteen AA batteries.

"It was a dopey little device," Sasson said.

"It was my baby."

◆   ◆   ◆   ◆

IN A WINDOWLESS conference room, cushioned chairs surrounded a long table in the center. Sasson was ready to demonstrate his prototype of the first digital camera to Kodak's upper management. He took a head-and-shoulder shot of one of the executives. Then he began to describe what he had done, cleverly trying to hide the twenty-three-second lag required to record each digital image on the magnetic cassette tape that stored them. The tape was then removed from the camera and placed in a purpose-built playback device that connected to a television. A black-and-white picture of the executive then appeared on the screen.

The people at the table were stunned. Some really loved the idea, and some hated it. Some were so shocked they said nothing. "The technical people were impressed that some stupid little kid in the lab could build this thing," Sasson recounted. But others launched a fusillade of questions and concerns. "Well, where would you store these images? You're not making a print. People love prints. People don't want to look at their pictures on a television set. Well, that image quality isn't good enough."

Sasson had no answers.

"It was like . . . shit!" Sasson told me. "I immediately wanted to pull back."

In hindsight, who could blame those critics? Kodak was,

after all, an institution anchored in Eastman's film photography. Here was Sasson showing pictures that didn't require film, photographic paper, or darkroom processing. This was a digital eruption in an analog world. "It was not a good way to get invited to the Christmas party," Sasson said. "The whole thing was too far out there to be seriously considered."

A colleague of Sasson told him privately, "Don't worry, the world will get there. They don't know it yet."

(2) THE PROTAGONIST in the 1951 movie *The Man in the White Suit* is a chemist played by Alec Guinness. His character invents a white suit that never gets dirty and turns out to be everlasting. This idea summarizes Martin Cooper's view on technology: it has to be durable and self-sufficient. Cooper invented Dynamic Adaptive Total Area Coverage—or DynaTAC. In English we call it the mobile phone or cell phone.

In the 1920s, Cooper's parents emigrated from Ukraine, where they had been subjected to persecution. His grandfather was the town butcher. He raised enough money to arrange a wagon train across Europe to Belgium. Cooper's parents then managed to move to Winnipeg in Canada and then to Illinois, where Cooper was born. He later studied engineering at the Illinois Institute of Technology.

Cooper is a slender man with curly, paper-white hair and beard. His career began at Motorola in the 1960s. "We were the most boring business in the world," Cooper quipped.

"When my mother asked me what I was doing, she'd have loved to hear me say 'I'm a doctor.' My mother knew who a doctor or a lawyer was. I used to say 'Well . . . I'm in the two-way radio business.' It broke her heart."

Since the late 1940s, Motorola—then known as the Galvin Manufacturing Corporation—had been a leader in car phones. In practice, they worked just as walkie-talkies do, but with a connection to the landline network supported by a switch-board operator. The car phones were convenient for people, but their performance was limited. Because there were only a small number of frequencies to begin with, the car telephone network was often congested and could handle only a few calls at any given time. Especially in big cities, as the installation and use of car telephones increased, frustrations among call-ers began to multiply as well. People had to wait a long time to be connected to calls from their car phones.

In 1968 the U.S. Federal Communications Commission opened up additional frequency ranges, expanding the possi-bilities for a "wireless" spectrum—meaning: "It will be pos-sible to make telephone calls while riding in a taxi, walking down the city's streets, sitting in a restaurant, or anywhere else a radio signal can reach," announced a Motorola press release. How? A geographic region was broken down into smaller seg-ments called "cells"—comparable to how ZIP codes and stan-dard time emerged from modular systems design. Within a single cell network, hundreds—if not thousands—of callers could share the spectrum at the same time. Once the person shifted from one cell to a different cell network—say, while

crossing a bridge—a set of computer-controlled radio trans-
mitters and receivers would keep the call connected. This signal transfer would be so automatic and dynamic that callers
wouldn't even be aware of it. The call would just drop if the
connection between the cell networks was bad.

This concept was the starting point for Cooper's DynaTAC.
In the early 1970s he approached his colleagues in Motorola's
industrial design group about converting his concept sketch
into a prototype. They offered very creative ideas. One idea
was a slider phone; another was a flip phone. Cooper selected
a single-piece design that looked like a brick. "The last thing
in the world we needed was the complexity of moving parts.
You know that moving parts are going to break," he said.

Cooper's team took about three months to prototype the
first generation of DynaTAC. The effort was informed mainly
by Cooper's previous experiences in building a commercial pager technology at Motorola. DynaTAC had thousands
of parts in it: radio, antenna, coils, capacitors, synthesizers,
oscillators, and batteries. As the prototype evolved, so did
Cooper's vision of a practical framework for modern mobile
telephony. "The phone by itself can't do anything," he told me.
"You needed a whole infrastructure around it." The idea bears
an uncanny resemblance to John Shepherd-Barron's vision for
the telematics of ATMs. Cooper kept shuffling and rearrang-
ing his ideas until he produced a proof of concept.

"When I get something working and demonstrate the principle, I lose interest," Cooper said. "I never thought that I was
a very good engineer," he added, "but what I'm really good at

is getting into the mind of the consumer. I'm the ultimate consumer myself." This point of view was the key to the success of his mobile phone concept. Updates to transmitter and receiver technologies, coupled with customer feedback, inspired further improvements to DynaTAC. The prototyping and refinement of the design continued, leading up to the demonstration of the first fully wireless call in 1973. At that time, Cooper's mobile phone weighed slightly more than 2 pounds. It had a battery life of thirty-five minutes. Hamstrung by regulatory constraints, commercial wireless service did not become a reality until a decade later.

Beneath the dense web of wireless communication lies an ascetic virtue that energized Cooper. If John Shepherd-Barron's idea for the ATM was fired by an inventive spark, Cooper's was more like a slow-burning ember: the belief that people are fundamentally mobile. Wires and cables only restrict them. "They infringe on our freedom, and to be free, you have to be disconnected. You have to be wireless," Cooper said. "If you're wireless, all of a sudden a whole bunch of things change. Things have to be small and light so you can carry them with you." This outlook influenced every bit of engineering that Cooper did.

◆　◆　◆　◆

MODERN MOBILE PHONES are hardly just phones. They are motifs of hyperfunctionality. Even the simplest phones are complicated. Developers and customers routinely become slaves to feature creep that bloats otherwise simple designs. This is where Cooper's wife, Arlene Harris, comes into the picture.

"I am unconstrained by the teachings of pure sciences," Harris asserts. "Marty, however, is far more methodical and grounded on the firmaments." Cooper first met Harris at a party in the late 1970s. Cooper was then a vice president at Motorola. "Her older brother never forgave me," Cooper said. "He wanted to tell me all his new ideas, and all I wanted was to talk to Arlene."

In recent years, Harris has been agonized by the mentality of mobile phone manufacturers who ignore the needs of senior citizens. "The phone companies didn't really care about that market," says Harris. She designed a mobile phone that was friendly to senior customers—really simple to use, and supported by an old-fashioned operator service. Her creation was the Jitterbug. Its features are basic. It has a numeric keypad, a screen with large-font display, and a big button to place a call.

Now consider this: Many people change phones every year. Each phone looks different, feels different, claims different improvements over its predecessor, and requires relentless software updates. Whether it's mobile phones or breakfast cereals, manufacturers push out "new" products for their own reasons. Within the business realities of "planned obsolescence"—or the technical realities of relentless improvements in operating systems, processors, and memory capacity—last year's optimally functional technology is rendered extinct, sent to the graveyard of gadgets.

By comparison, Tower Bridge, for example, has been safely functioning since 1894, requiring at most preventive upkeep and occasional fixes. The designs of the Jitterbug and the Tower Bridge uncover a question in the temporality of engineered systems: where is the line between transience and durability? The Jitterbug is a reminder of how difficult it is to create an intuitive, user-friendly interface. Simplicity is not about stripping down features to a bare minimum. It's about achieving elegance while maintaining performance. A technology is effective only if it enhances the user's value and experience. One can easily embed a miniature camera on a mobile phone screen for the sake of doing it, but only its usefulness can prove that it's a powerful tool, not a toy.

3 JUNE 1989. On the eve of the Tiananmen Square protests, Beijing cut off all communications with the outside world. Just two years earlier, in the United States, a handful of law enforcement officials and real estate agencies had purchased new digital image transmission systems from Kodak. This system—developed by Sasson and his team—was able to digitize and compress the images captured by a video camera and transmit them using a standard telephone line connection.

There was another early adopter: CBS News. It took Sasson and Kodak by surprise when CBS News used this technology to stream images from Tiananmen Square. "I was shocked. We had no idea," Sasson said while showing me an archival video in his home office, where hung a picture of him receiving the National Medal of Technology and Innovation from President Barack Obama.

At first, Kodak felt there was only a limited market for such a device, but Sasson viewed it as a critical step toward the development of commercial digital cameras. He needed to exploit image compression techniques to make the storage of megapixel images practical. Sasson's imagination now veered toward a digital camera with a built-in hard disk. By 1990, Sasson's camera had evolved into a very sophisticated prototype. Kodak engineers added JPEG-like image compression features before JPEGs were standardized. The camera had a memory card–like technology for storing images. It was a completely handheld unit. It had a color resolution of 1.2 megapixels. The simultaneous flowering of personal computers had opened up new possibilities. People could download the images onto their computers and do almost anything with them.

Sasson sensed sunny prospects. But soon he hit a wall. Kodak was rigid in its belief that digital cameras would cannibalize the company's lucrative film products. Sasson's innovation didn't mesh with Kodak's legacy. The company "basically told Sasson to take that box and go away; we don't ever want to see you again," a top executive from Kodak recalled recently. Sasson was frustrated. He left the digital camera business and looked for other opportunities. He even applied for a mission specialist job at NASA. No success.

A few years after he left Kodak, Sasson and his wife went

to Yellowstone National Park for a summer vacation. Like hundreds of other spectators, they were waiting for Old Faithful to erupt. As the geyser went off, Sasson looked around. All he could see was people taking pictures using digital cameras.

"It's happening," Sasson murmured to his wife.

"What?"

That was the moment when Sasson told his wife that he had invented the digital camera.

4 PIERS SHEPPERD takes his instincts seriously. He swings seamlessly among the worlds of arts, entertainment, and engineering as he runs a London firm that offers technical expertise for high-profile performance events like the opening ceremonies of the Olympics. "We realize the impossible. We make the exceptional. We build WOW," says his company web page.

When asked about his prototyping process, Shepperd declared that much of his job involves asking stupid questions. "I don't care about the details at all until I understand what the end point is," Shepperd said. "Even long before I get to the engineering, I spend quite a lot of time understanding what the client wants."

It gets tricky for Shepperd when his clients don't know what they want. "Some clients will give me a 3-D AutoCAD model, and they may have a very good spatial understanding of what it is that they're after. Other clients use only words.

Some even give you a piece of wood or a sketch from their children saying 'OK, this is what I imagine,'" he explained.

His clients present visions that are often vague and unrealistic. Shepperd's challenge is then to introduce structure—first in his own mind and then in his client's mind. "You're trying to put them in a space," Shepperd said. At the same time, "there's a part of my mind that says I need to think like an artist to permit self-expression." Artistic people assign great value to an object or an idea, and what it may mean to an audience. "It isn't until we actually build the object as a prototype that we learn if the significance is justified," Shepperd added.

Consider this hurdle that Shepperd faced during the planning of the 2012 London Olympics, for which he was the chief technical director: Film director Danny Boyle (the event's chief artistic director) and his team had a vision for an inspiring opening-ceremony scene depicting the industrial revolution. Their concept included elements like chimneys, steam engines, and looms. All of the proposed elements in the scenery were to be delivered live and full-scale in front of the audience within a total allotted performance time of ten minutes.

Prior to this scene was an act featuring the bucolic English countryside—"green and pleasant land"—set with real meadows, grazing animals, and water mills. It was technically unfeasible to install the industrial revolution scenery within the countryside set up in advance. The transition between the two scenes had to be natural, smooth, and seamless. Shepperd encountered another practical challenge. How and where would the large hardware—like the chimneys—be stored when not in use before and after the scene? Trying to find a place for them while not affecting other items—especially the Olympic cauldron appearing later in the show—proved to be almost impossible.

Shepperd tried negotiating with the artistic team. Could the chimneys be digitally projected instead of being built as full-scale models? Another possibility was to have two-dimensional fabric rolls with chimney images that could be quickly pulled up for the scene. These would eliminate the transition challenges that concerned Shepperd. The artistic team didn't budge. They were firm on having real 3-D objects that could be deployed in minutes. Moreover, for the whole scene to be effective, they needed at least ten chimneys.

Under these constraints, Shepperd's team started to work on numerous software and hardware models for the chimney. Then emerged what seemed to be a tangible method. They fabricated a series of concentric plastic rings that could be pulled by a cable from the ground using an aerial winch. The rings stacked within each other and were almost invisible on stage. But the testing showed that they were vulnerable to wind. Further, the scene didn't look very exciting when deployed.

At a time when he was desperate for alternatives, Shepperd somewhat randomly spotted the use of inflatable puppets for the Mary Poppins scene. It clicked. Shepperd could imagine an inflatable chimney. In practice, this was an ingenious method because it would require only a small winch inside the inflatable envelope that could deploy the chimney more gracefully.

The artistic team couldn't quite see the full potential— probably because the tower at that point didn't look much like a chimney—but Shepperd wasn't discouraged. He thought the inflatable approach was worth investigating further. There followed a long period of prototyping to develop a more real- istic chimney. A few months later, Shepperd demonstrated to the artistic team a full-height inflatable chimney supported by a winch cable from the top. His team had cleverly printed a brick-pattern fabric for the tower to make it look like a chimney.

Shepperd and his fellow engineers then worked robustly to develop the internal winch system and the blowers capable of keeping the chimneys looking solid as they lifted into the air. They added some foam rubber finishes near the base of the chimney to create what looked like a more substantial brick base. For an aesthetic touch, they also added a small smoke generator near the top so that the chimney could have wisps of smoke rising from it. Shepperd also deployed an aerial art- ist alongside the chimney as it rose into the air. "This made the chimneys more realistic, as the artist could pretend to be working on the masonry," Shepperd said. "It also allowed the camera to really understand the scale of what was being deployed."

This process of quick iterative design is really important for theatrical effects because it enables the whole team—both technical and artistic—to understand whether the end result will be worth the budget and effort. "Some things look and sound very good when described in a model, but are underwhelming when delivered at full scale," Shepperd pointed out. "The deployment of the chimneys ended up being one of the key iconic moments for the opening ceremony."

For Shepperd, the engineering process is different every time. His goal of providing mesmerizing multisensory experience is orchestrated with stopwatch precision. As with the case of the chimneys, Shepperd's basic design consideration is whether the structures need to *be* real or *feel* real? For all this work, rain, wind, and gravity provide Shepperd the constraints.

These challenges are different from building a bridge or an airport—projects that have firm, frozen objectives and specifications and rely on a cornucopia of standard case studies and learned experience. There's a well-defined design journey to prevent failure or liability. "If technology fails, man is the backup option," Shepperd said. To come up with something novel that triggers an emotional effect involves a lot of trial and error—and often far more error.

5 TWEAKING AND PROTOTYPING are basic human habits. Anyone who cooks knows a little bit about both. They are also powerful professional tools employed by engineers. Steve Sasson's work was founded on a steady, step-

wise stream of tweaks. The successive refinements were carried out with the knowledge that there's always another layer of onion to peel. It was an exercise in *functional prototyping*.

The digital-camera prototype was an important crutch for Sasson. He could show the prototype to his company executives without having to explain much of the technical details. They could look at it. They could touch and feel it. They could see their instant photograph on a television screen. As Sasson was challenged in that conference room, the spirit of the discussion was about a world of new possibilities—even as his colleagues were completely unprepared for the digital revolution.

With the digital camera, Sasson had no formal, planned specifications. No requirements were set in place. He was doing what we could call blind engineering: the final outcome was unknown. He was searching a wide range of solutions and didn't start off knowing that the charge-coupled device could lead to a digital imaging system. Nobody was complaining about film-based photography, nor did Sasson start out with the idea of building a digital camera. He was exploring an application for a new technology that his supervisor thought was interesting. For him the notion of the digital camera was as simple as "why not?" It was as if he was baking a cake, and making up the recipe as he went.

Prototypes are easier to react to. As Martin Cooper put it, "If my wife shows me a dress and asks, 'What do you think of this?' I can't tell much until she puts it on." Cooper's invention of the mobile phone emerged from *conceptual prototyping*. It started out with a specific vision. It required forward reasoning. Like a sculptor, Cooper had to chisel out the concept of DynaTAC while getting rid of useless elements and filtering out fragile ideas.

This design concept is also integral to Piers Shepperd's line of work. In *aesthetic prototyping*, functionality of the product becomes almost secondary to the sensory effect it will have on the viewer. When you see the Empire State Building, what does it make you feel? Does it fill you with a sense of wonder? Does it *impress* you? Then come the technical considerations. Can the prototype of the amazing skyscraper be 1 foot tall or does it need to be 80 feet?

◆　◆　◆　◆

THE CONNECTIVE TISSUE between Sasson's functional prototyping, Cooper's conceptual prototyping, and Shepperd's aesthetic prototyping is the principle of *test-driven development*. Testing relies on data. Testing also generates data. But data are not always available to support design decisions. Nevertheless, these decisions need to be made. That's probably why engineers rely on prototypes as reasonable substitutes for data.

In the case of Sasson and Cooper, people might have thought about the concept of a digital camera or a mobile phone, but no one had effectively prototyped one before. These technological changes represent *transformation*—systems built from existing tools in the same way that the genetic characteristics of bacteria are shaped through the assimilation of DNA supplied by the environment. These changes also confirm the broader fact that there is no "perfect" design. As the Japanese notion of *wabi-sabi* makes clear, everything is imperfect, everything is impermanent, and everything has room for improvement.

Prototypes are also helpful in highlighting the potential threat of design fixation. Locking in on a path prematurely begins to thwart innovation. Risk aversion creeps in. Psychologists refer to this phenomenon as the trap of *Einstellung*—a set bias that impedes better solutions by staying glued to knowledge that's familiar and favored, or adhering to an experienced frame of reference. This outlook initially hindered Kodak's business strategy—just as it blinded Vallière from recognizing the explosive potential of Gribeauval's agile, lightweight cannons. Similarly, Kodak's fixation with film photography crippled its capacity to see the power and prospects of Sasson's digital-camera technology.

Prototypes create new capabilities. Prototypes foster adaptation to new forms, new expectations, and new offshoots of technologies. Prototypes are the starting points toward our ultimate creative destinations. If you consider steam power as a prototype application, it took about 120 years—that is, four working lifetimes—to reach a saturation point. From the days of the earliest mechanical (verge escapement) clocks, the errors of accuracy in clocks have diminished by six orders of magnitude in six centuries. At one point in their evolution, clocks used to be off by thirty minutes a day, but now their error rate is down to a fraction of a second. Land and air transportation technologies have evolved similarly.

Technological performance seems to double roughly once every thirty years until the technologies themselves become saturated or obsolete. Estimates suggest that efficiencies of some technologies have improved by fourfold to eightfold if not more during every generation since 1840. The performance of powered balloons has risen remarkably from the very sluggish rates of the late nineteenth century, achieving an almost tenfold improvement in a single generation to reach the current levels of commercial and space transport. In the case of Sasson's digital camera, it took twenty-odd years for the resolutions to go up from 0.01 megapixel in black and white to 1.2 megapixels in color—a 120-fold improvement. More recent camera models are emerging with astounding capacities, including superior lenses, digital and optical zooms, and high-definition video resolution. These dramatic improvements have occurred in just the past five to ten years. Other systems—for example, the capacity of the telecommunication spectrum supporting mobile phone networks or the density of the semiconductor chips—have peaked at exponential rates.

These increases are a "strange human rhythm," says John Lienhard, a cultural historian of engineering. New technologies are now created in months, sometimes weeks. "It's unconscious. It's inexorable. The inventive mind seems to be animal instinct as much as volition," Lienhard observes. "Invention flows from our inner being. It is a powerful river that cannot be dammed or deflected . . . it's the way we insist on life. Invention is the primary means by which we rage against equilibrium and death."

(6) IN THE AUTUMN of 2009, some of the world's prominent leaders in science, technology, and business gathered at an awards ceremony in London. It was a black-tie gala organized by the *Economist*. Two shy engineers attended this event. They had never met before. They grew up and worked in very different circumstances. On the surface, their works were unconnected. But their efforts culminated in a fusion that probably neither of them predicted.

When they greeted each other for the first time, others took pictures of them with their mobile phone cameras. Their names were Steve Sasson and Martin Cooper.

Eight

LEARNING
FROM OTHERS

(1) IT'S NOT ALWAYS EASY to please an engineer. Victor Mills hated washing cloth nappies. At sixty, he did that often when he had to take care of his granddaughter. He thought it was "a mess."

Mills was a chemical engineer. He worked for Procter & Gamble for over thirty-five years, where his résumé included several notable achievements. He devised a clever process to stop the separation of oil in peanut butter; people loved Jif. He found a way to eliminate the clumps in Duncan Hines moist cake mixes; smoother cakes are always a hit. He developed a system to uniformly stack and pack potato wafers; Pringles was a sensational success.

In the late 1950s, P&G was trying to figure out how to make

the best use of a recently acquired pulp mill—Charmin Paper Company. At home, Mills was frustrated with cloth nappies. At work, Mills was vexed about the pulp mill. Mills put two and two together: use the mill to produce absorbent paper for nappies.

Mills enlisted his staff engineer Robert Duncan to help produce a test model. They stacked thick, rectangular pads of paper and used polymer wrapping for stable inner and outer layers to withstand wetness. For the actual testing, they used Betsy Wetsy—a doll that squirted water—as proxy for a urinating child. Mills's grandchildren offered an opportunity for more nappy testing during long road trips.

The result was a sensational consumer product: Pampers. As the world's first successful disposable nappy, Pampers gave birth to a multi-billion-dollar child care products industry. An array of health products then emerged to improve hygiene and convenience.

Mills's group was not the first to think of throwaway nappies. What made their effort worthwhile—and ultimately successful—was the customer feedback they applied to refine their product. Mills's team consulted with mothers, pediatricians, economists, and environmentalists.

Norma Baker, a home economist and Mills's colleague, was skilled in the art of connecting with customers. She advised Mills to create two nappy models: a tape-on version and a pin-on version. Both turned out to be more effective than plastic pants, especially on warm days in places without air-conditioning. She reported that mothers found P&G nappies too expensive—a whopping ten cents apiece. Mills found a way to mass-produce them at lower cost. Until then they had been hand-sewn on the order of tens of thousands. Baker's field studies made it clear that nappies also needed to be extra absorbent. Subsequently, a new market for disposable nappies materialized: patients in hospitals.

The story of disposable nappies is a lesson in *responsive design*. Engineers worked with the end users to improve their product. It's an example of how social interactions with customers can be a powerful force in influencing engineering design. The same dynamic was at work in the story of the ketchup bottle.

♦ ♦ ♦ ♦

In 1869, around the same time the First Transcontinental Railroad project in the United States was completed in Utah, the H. J. Heinz Company introduced the clear-glass-bottle design for ketchup. The glass container had obvious advantages: it was easy to manufacture, and people could clearly see how much ketchup was left.

The biggest hassle, though, was getting the ketchup to flow out of the bottle easily. It required shaking, poking, knifing, and, above all, persistence. Then there was the problem of syneresis—the separation of watery tomato serum in the bottle that was called the "ketchup spit"—which was unappetizing. Struggling to come up with a solution, Heinz decided to learn from its consumers. The company's research with six- to twelve-year-olds showed that kids liked to draw ketchup art on their plates. Heinz found the idea inspirational.

Corporate engineers and designers spent days with their modeling software and arguing about trade-offs. The upside-down ketchup container made out of polyethylene terephthalate (PET) became a reality. An orifice controlled the ketchup flow, and the container had a stable molded hinge. Poking was eliminated. Every squeeze squirted a single serving of ketchup. Knifing and shaking were gone. A trap cap to pneumatically "suck back" the watery stuff was put in place. The look and feel of ketchup dramatically improved. By applying a little extra pressure on the PET container, kids could draw a smiling sun on their plate. Artistic possibilities were now boundless. The container even emptied "as fast and cleanly as a gas can at a NASCAR pit stop," as one journalist put it. Consumer experience was elevated.

The idea of using the valve and self-sealing cap design for ketchup was hardly novel. Many shampoo bottles work the same way. But Heinz's approach to improving its product design—notably the ergonomics of comfort, grip, softness, and squeeze—was informed directly by the end users themselves. When it relates to the engineering of consumer products, education always goes two ways, even if it's a problem that consumers don't lose sleep over.

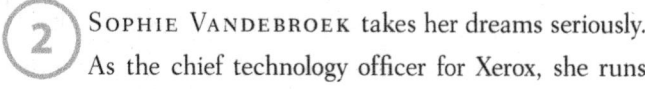 SOPHIE VANDEBROEK takes her dreams seriously. As the chief technology officer for Xerox, she runs

"dreaming sessions." These are like improv sessions with end users—or *coinnovating*, as Vandebroek calls it—so Xerox's solutions and services end up being useful to their clients.

Vandebroek is originally from Leuven, a township near Brussels in Belgium. "I was one of only fifteen girls in my class of five hundred students," she said. "It was the most difficult bachelor's degree program to get into, and also the only one for which there was a separate entrance exam." After years in the technology industry, Vandebroek began to develop a unique respect for social scientists. "They are trained to really observe and deeply understand human behavior and processes. They are able to articulate what would really make a huge difference in the customers' lives," she added.

Vandebroek couldn't expect this sort of deep understanding of human behavior from her fellow engineers. This deficit on the part of engineers became evident when Xerox launched color copiers decades before color printers became ubiquitous. Making a copy of a color original was a breakthrough idea from an engineering perspective, but Vandebroek said the idea didn't make any sense at that time. Why? There were relatively few original color prints to begin with. "This clearly wasn't the pain point of the customer. Although a technological breakthrough, it didn't sell. It was way too early."

Something similar happened with the first color digital production press in the 1990s. Xerox engineers packed in many sophisticated work-flow tools and an automatic color calibration software system that they thought would be a blockbuster in the world of offset printing (a widely used process of mechanical ink transfer from a printing plate to a rubber roller and then onto paper). The digital press also seemed a clear vehicle for saving time and money. "When people actually bought these machines, they didn't use any of these automated features and capabilities we had built in, and they continued to do color calibration the same old way it was done in analog offset presses," Vandebroek said. "Only after our ethnographic scientists went and observed our clients in their shops did we engineers realize what was happening."

Experiences like these were not completely new for Xerox. In the early 1980s, Lucy Suchman—then a staff researcher specializing in how humans interact with new technologies— videotaped some top scientists and engineers trying to make double-sided copies. They struggled. The photocopier was way too complicated.

The digital production press had the same problem. Were it not for anthropologists and ethnographers, Xerox engineers wouldn't have known why their products weren't being used in the way they were supposed to be used. Xerox anthropologists began "shadowing" users. During these ethnographic studies, the offset-printing vendors piled on their dislikes and complaints. In response, all it took on the engineering side were some software tweaks. The machine's operations were simplified, and the clients were pleased.

It was a "bingo" moment that engineers couldn't have recognized without the help of anthropologists. Unfortunately, this realization occurred only after product launch. Executives at Xerox were getting nervous because the users were complaining. "This notion of really understanding your client in the very early stages of research was definitely counter-intuitive, and just very, very powerful," Vandebroek said. "We now involve the clients from the very early stages and have ethnographic experts participate in all big research projects. The client's environment is our laboratory."

The crux of the issue may not be the absolute complexity of a photocopier or a printer, but how the customer *perceives* the complexity of the technology, as Suchman explains in her book *Human-Machine Reconfigurations*. She raises the notion of *contingent coproduction*—which is similar to Vandebroek's dreaming sessions—or the idea of codesigning with the end users to make technology friendly and intuitive.

Smartphones, GPS, microwave ovens, cars—they're all incredibly sophisticated and complex technologies, but making them approachable and accessible in collaboration with customers is a hallmark of engineering innovation. After all, it's easier for engineers to learn about their customers' pain points than to expect their customers to learn software routines.

3 SONY'S TAPE RECORDER was having an existential crisis. After years of hard work, Sony found out that no one was really looking for a tape recorder. "The tape recorder was so new to Japan that almost no one knew what a tape recorder was, and most of the people who did know could not see why they should buy one," Sony's legendary chairman

Akio Morita recounted in his memoir *Made in Japan*. "It was not something people felt they needed. We could not sell it."

Morita was discouraged that people were willing to spend more on vanity art than on something like a tape recorder that might have a deep practical value. It was a crafty solution in search of an opportunity until a group of overworked Japanese court stenographers adopted it. They were an enlightened group of users. For them the tape recorder was a godsend; "to them it was no toy."

The need for a new technology—like improved nappies, plastic ketchup containers, color copiers, or tape recorders—is not always intuitive. The value of a consumer appliance is short-lived if it's simply about the technology. Social perceptions drive technology adoption and acceptance in society. Countless companies and their products share a version of the journey made by P&G, Heinz, Xerox, and Sony.

Refrigerators are a good example. Besides the obvious merit of food preservation, what truly seemed to influence the sales of refrigerators was social marketing. When refrigerators were first introduced in the 1920s, from a narrative standpoint it was an easy sell: they "preserved women" by aiding the servantless wife. From a technology standpoint though, we know how refrigerators have lengthened the preservation of perishable goods and stimulated the explosive growth of supermarkets and their supply chain systems. Washing machines have a similar pedigree. The initial claim was that they saved women from the household drudgery of doing laundry. Soon what were once "prestige products" came to be valued in the same utilitarian spirit as cutlery.

◆ ◆ ◆ ◆

UNDERSTANDING CUSTOMER HABITS is a treasure chest for new product development. If Xerox's approach has been to rely on the wisdom of the social sciences, then Toyota's strategy comes from a different angle. To better understand its customer preferences, Toyota created a "set of planning and communication routines" to "market the goods that customers want to purchase and will continue to purchase," according to a classic *Harvard Business Review* article entitled "The House of Quality."

Using this framework, dubbed the "quality function deployment," Toyota "improved its rust prevention record from one of the worst in the world to one of the best by coordinating design and production decisions to focus on this customer concern." In a practical sense, Toyota's product engineers and designers took a modular systems approach by breaking down customer concerns into fifty-three key items over eight different design levels, "covering everything from climate to modes of operation. They obtained customer evaluations and ran experiments on nearly every detail of production," the article explains. Toyota could well have treated customer feedback as a "soft" constraint, but this example illustrates that customer concerns were not regarded as optional issues but became an integral part of Toyota's design decisions to handle rust.

Similarly, Toyota engineered the second generation of the Avalon, its flagship vehicle, to eliminate what the company called the "stone-pecking noise"—something that was caused by the deflection of pebbles from the tires onto the chassis, and that customers complained about. Where others might have disregarded the problem as a trivial issue, Toyota undertook a total systems analysis to eliminate the annoying noise. In doing so—as an article in the *Los Angeles Sentinel* reported—they started "exploring new multilayer floor-carpet silencer," and "engineers added sound deadening felt to the luggage compartment, revised door seals and thickened the side window glass, plugged dashboard holes, selected new tires, and repositioned the windshield and wipers."

From Toyota's perspective, the customers and their preferences appear to be the constraints, and the design issues are the trade-offs. Process innovations stemming from Toyota's quality control principles are not about creating cool geometric models, in the view of Michael Kennedy, a consultant in lean production engineering; rather, it's thinking about the "range of customer interests, what you can do to meet them, and where you have gaps in your knowledge."

Learning from others' experiences shouldn't be an accessory to engineering design; it should be a core technical necessity—just as music isn't successful if it stays in the mind of the composer, but only when it spreads across the outer world to make listeners sing, dance, feel, and fall in love with it. "The biggest difficulty is moving a new technology into people's lives. Once the public recognizes the advantage this technology brings, they come to expect it," writes Morita. "What housewife would want to go back to the washboard?"

(4) IN THE SUMMER of 1853, U.S. president Millard Fillmore commissioned naval commodore Matthew Perry to travel to Japan so that the two countries could become trading partners. Japan was isolationist at the time. Opening up to the Western world was a shocking prospect. Perry and his fleet showed up on the shores of Edo Bay in four well-armed black ships. After prolonged negotiations at the 1854 Convention of Kanagawa, the Americans and Japanese came to an agreement. Japan was "liberated." It moved into a new era of trade and development.

Referring to the vessels that Perry used to awaken Japan, historians use the term "black ship effect." The engineering profession also needs a black ship effect. Engineers should rise above the comforts of cold, mechanistic, isolated problem solving. The best people to help the engineering profession with this creative liberation are cultural anthropologists.

The difference between a typical engineer's approach and an anthropologist's approach is simple. Engineers naturally tend to focus first on the product and then on its users. For an anthropologist, it's the exact reverse: people come first, then the product. "It's really about shadowing the people because ultimately for anything to be successful, it has to be user centered," says Margaret Szymanski, an anthropologist at Xerox's Palo Alto Research Center (Xerox PARC).

Social sciences like anthropology are best known for their "unbounded inquiry," writes industrial anthropologist Francisco Aguilera. He stresses that anthropology is not about describing the forests and the trees, but about "pursuing the ecology into the bordering grassland." But unfortunately, branches of cultural anthropology—like ethnography—have historically been overlooked by other professions, especially engineering. Many engineers regrettably think that social sciences are "common sense," a sentiment that the late Diana Forsythe, an eminent anthropologist of technology and computing, had criticized as the "problem of perspective"—the difference between what engineers know and what they *assume* they know about customer preferences. Another advantage of working closely with anthropologists is the ability to appreciate the "problem of order," as Forsythe described; to rely not on snapshot feedback to make rapid tweaks, but to continue the process of social observation over time to build effective, long-lasting products.

As a human tendency, engineers sometimes take pride in perfectionism, to the extent that they overengineer product designs and worsen the customer experience. Even after all these years, how many of us are still able to effortlessly open clamshell plastic packages without scissors—or even with scissors? And why can't we unwrap those modestly sized airplane snack crackers without crushing them? There are clearly limits to how engineers sometimes think.

In *Toyota Culture*, Mamie Warrick, an administrator at the University of Toyota, tells the story of engineers involved in the production of the RAV4. These SUVs didn't have any cup holders when they were first launched in the U.S. market in the mid-1990s. As Warrick describes,

> To help the chief engineer understand the situation, one of our distributor members picked up the chief engineer in the current RAV4 model, took him to the local 7-11 and bought the guy a 32 ounce hot cup of coffee. The purpose, of course, was for him to discover there was no place to put that cup. So the American team member helps the chief engineer into the car and gives him the cup of coffee. The chief engineer is so delighted with the coffee, he doesn't bother to put it down and just downs it. It's steaming hot! (The Japanese have a higher tolerance for hot liquids.) And at that point he has the empty cup and realizes, ah! There is no place to put it, there is no cup holder. And then the point is made.

Whether we're talking about massive systems development or conservation efforts, the engineering profession must go beyond its traditional analytical trappings and embrace disciplines like cultural anthropology as partners for better understanding society. The wisdom of anthropology can help engineering take a more enlightened approach in appreciating our interdependencies. It's only at the intersection of numerous disciplines—far beyond our cushy comfort zones—that innovations catch fire and spread.

◆   ◆   ◆   ◆

MUSIC AND MOTION PICTURES offer fine examples of the art of collaborating with audiences. The audience is very much included in the creation process, says Rob Cook, an Academy Award–winning Pixar engineer. "No one has precircumscribed the set of things you can create. In engineering it's easy to get far afield from creating something that's useful and has an impact. You can have an idea in your mind, 'Oh, this is exactly what would be useful to this particular type of customer,' but then you go build it, and it turns out 'No, actually, it doesn't really work for them.'"

Tastes differ, so an average of customer preferences would mislead. There's no formula to get around this preference paradox. If Henry Ford had done a customer survey, he might have received requests for faster horses. Consider air bags, a valuable safety feature in modern cars. How many of us could have thought of a balloon blowing up from the steering wheel during a crash as a design concept? Despite these intuitions, engineers would be better off learning to listen at a deeper level to what people are trying to say, which is different from producing a checklist of features requested by users. "If you really deeply understand the sorts of things people are doing and create something in response to that, that's when it really strikes a chord," Cook said.

The late Steve Jobs, who led Pixar, related the emotional elements that influence the forms and formats of technologies. In an interview for *Fortune* magazine, he said: "We don't have good language to talk about this kind of thing. In most people's vocabularies, design means veneer. It's interior decorating. It's the fabric of the curtains and the sofa. But to me, nothing could be further from the meaning of design." Using the iMac computer fan as an example, he continued:

I was adamant that we get rid of the fan, because it is much more pleasant to work on a computer that doesn't drone all the time. That was not just "Steve's decision" to pull out the fan; it required an enormous engineering effort to figure out how to manage power better and do a better job of thermal conduction through the machine. That is the furthest thing from veneer. It was at the core of the product the day we started. . . . This is what customers pay us for—to sweat all these details so it's easy and pleasant for them to use our computers. We're supposed to be really good at this. That doesn't mean we don't listen to customers, but it's hard for them to tell you what they want when they've never seen anything remotely like it. Take desktop video editing. I never got one request from someone who wanted to edit movies on his computer. Yet now that people see it, they say, " 'Oh my God, that's great!' "

The essence of a good technology is that it's intuitive, and it evolves. Ideally, you don't even want to know it's there. Many modern interactive technologies have become so instinctive that children learn how to flick, pinch, and zoom on their tablet screens even before they learn how to walk, talk, or write. As Marissa Mayer, the CEO of Yahoo! points out, "All of the complication lies underneath, just like an iceberg, there's just that thin little layer that you interact with." But the curious thing is that many intuitive technologies we rely on would have never come about in a focus group.

5 — THE INDONESIAN ISLAND of Bali is renowned for its water temples. The Balinese doctrine of *Tri Hita Karana* takes a holistic view: God, nature, and humans are interrelated. Underlying the emerald-green rice terraces in Bali is an integrated system of organic farming and shared water management called *subak*. This cooperative practice among farmers has been governed by millennia of transcendental belief under the aegis of the upper-caste priests of the water temples.

As evidence of indigenous engineering design, the Balinese people built sophisticated irrigation tunnels dating back to the eighth century AD. These tunnels facilitated the water sharing between upstream and downstream farmers. The artificial ponds supporting the rice fields depended on seasonal monsoon rains. The rainwater that washed away from the volcanic rocks deposited phosphate in the pond, transferring some vital nutrients to the rice paddies.

Twice a year the harvests occurred in a well-coordinated fashion. This synchrony also had a powerful time-tested usefulness: it succeeded in controlling pests. After a harvest, all the pests would be wiped out for a time, which was a much better scenario than having pests year-round. Through this practice, "the water temple networks optimize the trade-off between pests and water," says Stephen Lansing, an anthropologist at the University of Arizona. By maximizing water conservation and simultaneously reducing pest attacks, Lansing adds, "one can see that these water temples play a useful role in finding the appropriate scale of coordination to optimize those two opposing constraints."

Everything seemed normal in this ecosystem—except when some government technocrats decided that this age-old process was inefficient. Officials involved in the green revolution—an outcome of agricultural engineering—managed to convince farmers to use their "technology packets" of high-yield varieties of seeds, pesticides, and chemical fertilizers. The goal was to substantially intensify rice production. The farmers could have four or five crop rotations per year in place of two. The notion of efficiency—treating a crop as an instrument to multiply output—uprooted the subak tradition, just as Gribeauval had stripped away from the French cannons all the artistic elements that he thought were meaningless.

After initial increases in yield, the Balinese results turned devastating. "Miracle rice produced miracle pests," notes Lansing, who has studied and documented these effects in his scholarly book *Priests and Programmers*. The IR8 variety of rice that was introduced turned out to be vulnerable to brown plant hoppers, leading to a loss of 2 million tons of rice in 1977. The upgraded IR50 strain succumbed to tungro virus.

Soil erosion and a huge disruption in water schedule ensued. This was a "litany of horrors," Lansing says. "We've made colossal mistakes. . . . These systems are collapsing right before us."

In the Bali case, the scientific community introduced a technology without respecting an already productive system rooted in rites and ancient tradition. Despite the disastrous effects of this shortsighted approach, government officials weren't convinced that the system overseen by the temple priests was any good. It took computer models to sway them. The results showed that the highly sophisticated pest aversion strategy of the traditional approach was far superior to the new technology.

♦   ♦   ♦   ♦

THE BALI STORY reminds us that ideas can have split personalities. In the 1960s, the green revolution transformed India's farmland from a begging bowl to a breadbasket. But the same concept backfired in Indonesia, and it may never get off the ground in parts of Africa. Cultural considerations are powerful determinants of a technology's success. Even more, nothing on earth has only benefits. Every positive thing can also have bad outcomes. That's why mindlessly privileging efficiency and productivity while not considering other native factors is a flawed approach. More efficiency may actually lead to more consumption. But complex social circuits make it difficult to predict when things that seem good will go bad.

Engineering is no exception. The same principle that's used

to produce a software security patch can create a destructive computer virus. The space program has its origins in the development of intercontinental ballistic missiles. Internal-combustion engines have helped humankind reach remote corners of the planet, but they are a major contributor to pollution and climate disruptions. Optimization algorithms have increased financial yields but also had an "invisible hand" in financial disasters. From the conveniences of packaged food to the casualties of processed food, engineering plays a central role. Location-enabled technologies—such as enhanced 911—can augment public security but fuel stalking. The cell phone technology was created to offer the freedom of mobility, but that freedom has also had a reverse effect: people are now tied to their jobs and the so-called "social networks" that blur the lines separating work, family, and everything else. Gone are the days of going online; many of us now *live* online.

Life as we know it is a string of choices that lead to consequences. Intended or not, consequences sometimes can't be realized until decades have elapsed. It's not always possible to foresee the real possibilities of our creations—something philosophers refer to as *designer fallacy*. The Chinese invented gunpowder centuries ago, but it was Europeans who applied the technology to power their cannons in the process of modernizing warfare.

There's also the *intentional fallacy*: design geared specifically toward villainy. Hitler's engineers found efficient ways to commit genocide. From designing "reliable" ovens to "optimizing" the quality of genes to "standardizing" the construction of concentration camps to "tracking" inmates like packages to "mass-producing" cadavers, these were morally repugnant applications of the engineering mind-set. As history has clearly demonstrated, these examples are instances in which engineering principles did, unfortunately, work in practice.

Is engineering good or bad? This is a "curate's egg" question. We should look instead at a continuum of risks and benefits, as we often do in matters of consequence. "Technology is not like a gentle rain that falls on all equally, as Buddha suggested in the parable of medicinal herbs," Levent Orman of Cornell University has written. "It is more like a thunderstorm that benefits some but devastates others."

Like it or not, the engineering mind-set is use oriented and outcome focused. It craves specific end points. Perhaps because of these core attributes, sociologists Diego Gambetta and Steffen Hertog controversially proposed that engineers and individuals with technical backgrounds are overrepresented in terrorist and other fundamentalist groups. It would be ridiculous, however, to link engineering and social radicalism. In fact, to the contrary, it's the engineering sense of being *oriented to the mission* that Gambetta and Hertog prominently discuss in their thesis. They conjecture that "engineering as a degree might be relatively more attractive to individuals seeking cognitive 'closure' and clear-cut answers as opposed to more open-ended sciences—a disposition which has been empirically linked to conservative political attitudes."

Although feeling certain about particular outcomes can motivate people toward antisocial behavior, mental illness is an important contributor too. One could argue that methodical systems-level thinking also plays a crucial role. "Asking how organized a subject is may be far more important than asking whether the subject is mentally ill or not," says Robert Fein, a forensic psychologist specializing in the prevention of targeted violence, including assassinations. Further, like all other people, terrorists need to have skills relevant to their horrible mission in order to be successful.

These complex social issues go deeper than traditional analyses and arguments. At a fundamental level, they may point toward tragic flaws that are an inherent part of being human. Our lives, beliefs, and experiences set us on specific paths. Certain people are predisposed to see life's challenges as hammers and nails, and such a leaning can't be blamed on the powers of engineering. The best we as a society can do is to periodically review our social contract with engineering. The best engineers can do in return is not to waver on civil responsibilities and the trust that society bestows on them.

(6) ENGINEERING HELPED put humans on the moon. Engineering has dramatically enhanced our living standards. But why haven't we been able to eliminate poverty and inequality? These "moon and the ghetto" problems, as economist Richard Nelson dubs them, exist because there are no clear paths to solutions. We don't have the know-how to effectively deal with a wide range of thorny issues. Backward thinking may become convoluted.

Many of our social challenges are fuzzy. They are bound-less, have poor structure, and lack an expiration date. Most important, these challenges impose uneven social costs that change with the pace and priorities of our societies. These costs could be either useful or harmful. For every technical solution, such as in the construction of a new road, there often ought to be a commensurate market-based solution, like peak-load pricing. A purely technical solution, without the market forces to support it, would be like blood flow with-out oxygen.

Engineering can help tackle many but not all social chal-lenges. Moreover, we will continue to encounter new oppor-tunities and challenges in developing complex systems that engineering alone will not be able to answer. Information and communication technologies have already started producing new kinds of relationships between humans and engineering creations, which in turn are producing new kinds of social norms and interactions. Only through improved understand-ing of the subtleties in human behavior can engineering continue to boost our economies and serve our society. To expand its peripheral vision, engineering must be educated and enriched by the vision, wisdom, and inspiration of cre-ative arts, literature, humanities, sciences, and philosophy.

Technical education shouldn't reinforce a commodity men-tality, but nurture a collaborative mentality. In doing so, the engineering profession needs to embrace and apply fresh forms of aesthetics, a lively sense of openness, and an energetic type of pluralism. Engineering can bolster its performance and endurance by learning new sensitivities, by capitalizing on new synergies, by better relating to social sensibilities, and by continuing to adapt to cultural necessities.

Ultimately, what matters are not first impressions, but lasting impressions.

Fade-Out

A MIND-SET
FOR THE
MULTITUDES

(1) On May 16, 1961, a British engineer working in
Hollywood was reading the *New York Times*. A rotund
man in his early sixties with a sagging double chin, he spotted
an odd story: "Man in Bush Wants Birds Kept in Hand." It
was about an Australian man whose suburban house had
been attacked by kookaburras. These birds had "peppered
hundreds of holes . . . with their beaks, making a noise each
time 'like someone hitting a drum,'" the article reported.

On May 24 that year, the *Los Angeles Examiner* ran a piece
about another strange incident: "Man's Face Badly Slashed
by Owl." Months later, thousands of sooty shearwaters went
berserk and invaded Capitola, a coastal community on the
Monterey Bay. The newspaper headlines screamed: BIRDS
BLOCK TRAFFIC, CITY DEEP IN FEATHERS, DYING BIRDS JAM

SANTA CRUZ. "The place was black with them," a police officer said. The shearwaters are usually the docile type, but they were "crying like babies" and "smashed into cars, apparently attracted by headlights."

The engineer was intrigued. He liked to watch birds. To study their erratic behavior, he rented 16-millimeter films like *Western Birds at Home*, *Birds of the Countryside*, *Birds of Prey*, *Feathered Beachcomber*, *Birds of the Dooryard*, and *Journeys on Wing*. He then delved into a novelette published by Daphne du Maurier in *Good Housekeeping* magazine. It was about predatory birds marauding a seaside town.

He was captivated by the possibilities. Sipping his crème de menthe, and puffing on his foot-long perfecto, he started to jot down ideas on foolscap paper for his next motion picture. He called it *The Birds*.

The engineer was Alfred Hitchcock.

(2) THE ENGINEERING MIND-SET can be applied successfully in every walk of life because its core elements (structure, constraints, trade-offs) and its basic concepts (including recombination, optimization, efficiency, and prototyping) are equally effective in finding solutions to nonengineering challenges. We can see all these aspects converging clearly in the work of one of the most famous film directors of all time, who studied and had a "thorough grounding" in engineering. His early technical education had an important influence on his creations.

Alfred Hitchcock's goal was to convert a dream into something as real as a clock's ticktock, so his audiences would be "looking at a nightmare." Everything for Hitchcock was rooted in technical logic, even creating a suspenseful cinematic moment—the immediacy and very essence of the experience that would be thrilling and "chilling movie audiences long before air conditioning."

Just as a skyscraper's stability depends on the strength of its joints, Hitchcock's storyboarding relied on the precision of camera angles. He was a master of montage—the epitome of the modular systems approach in cinematic editing—which has been supremely influential in modern cinema. Hitchcock's target was his viewers' nerve endings; he wanted them to feel as though they were "dipping their toes in the cold waters of fear."

Rear Window, with Jimmy Stewart, is a great example of Hitchcock's modular thinking, and his operation within the notion of structure, constraints, and trade-offs. There's "a man in one position for the whole picture," Hitchcock once explained. "He never moves. Yet you have a close-up of Mr. Stewart. He looks, and you cut to what he sees, and you cut back to his reaction. And by the use of visual means you create ideas in his mind."

"And to show you how flexible the medium is, let us assume that you have a close-up of Mr. Stewart. He looks, and we cut to a woman nursing a baby. Cut back to Mr. Stewart. He smiles. Now what is Mr. Stewart? He's a benign gentleman. Take away the middle piece of film, have both the close-ups—the look and the response—and insert a shot of a girl in a bikini. He looks, girl in bikini, he smiles. Now he's a dirty old man."

Hitchcock's later movie *Psycho* reconfirmed the filmmaker as the Scheherazade of suspense and showmanship. *Psycho* featured Hitchcock's most inventive scene—arguably one of the most famous in motion pictures—in which a nude woman in the shower is stabbed to death. But "cinematically, there wasn't a single shot of a knife touching [the] body anywhere," Hitchcock said. "It was completely illusional."

It was the magic of his scissors that catapulted *Psycho* to its greatness. It was a spectacular exercise in modular systems engineering. Cutting "isn't exactly cutting. Cutting implies severing something. It really should be called assembly," Hitchcock once explained, adding that the "assembly of pieces of film which moved in rapid succession before the eyes create an idea." Hitchcock's approach as a technical artist was completely different.

"You could not take the camera and just show a nude woman being stabbed to death. It had to be done impressionistically," he explained. Hitchcock was able to create fear by mixing and matching short segments of film—seventy-eight frantic fragments of the knife, face, hands, shower, feet, water in the tub, and the dark shadow on the curtain. It was the unreal mirroring the real. The entire scene lasted about forty-five seconds.

With *Psycho*, "as the film went on there was less and less violence, but the tension in the mind of the viewer was increased considerably—it was transferred from film into their minds," Hitchcock said. This point suggests that he and Clarence Saunders of Piggly Wiggly weren't much different in their philosophies: they primed and anchored their audiences, letting them do all the work. "Towards the end [of the movie] I had no violence at all but the audience was screaming in agony," Hitchcock said, "thank goodness."

(3) LIFE PRESENTS US all with vexing challenges. As citizens of high-performance cultures, we are expected to engineer useful choices. In an era when ideas and financial resources seem as permanent as a fizzle in a freshly opened can of soda, how can we make the most (and better) out of less? How do we tackle the broader inefficiencies that metastasize inside our economies, education, health care, and governance?

Intrinsic in these challenges are elements that engineers deal with every day. In *Think Like an Engineer* we have met a procession of engineers who have applied the power of modular systems thinking and backward design to produce solution spaces. We've seen that engineers come in extraordinary varieties, and that the engineering mind-set is more a smoothie than a mixed fruit salad. Its ingredients aren't always so obvious. Furthermore, we've learned about how people are born, adopted, or married into the engineering profession in myriad ways. There are just as many ways to practice engineering as there are to achieve inner peace and harmony. We've explored how engineers continually learn from failures, successes, and interdependencies to refine their approaches. Clearly, improvements in efficiency need optimization, which in turn benefits from prototyping. We've also considered how engineers capitalize on the triumvirate of structure, constraints, and trade-offs to create technical revolutions.

The engineering mind-set is not a panacea; it's an enduring cognitive archetype and a durable practical construct for life. As a higher form of consciousness, engineering with its "blind alleys and disappointments [is] still very much a story of moving forward," the Smithsonian historian Tom Crouch affirms. "Even from mistakes you discover something that's useful in moving you forward." There's hardly a scenario in which the engineering frame of mind isn't beneficial. In the same vein, it's an alluring impulse to dream of a world filled with engineers. But why? Even if the dream of mass-producing the engineering quality of mind were to come true, would we be able to solve all our challenges? I doubt it. Crucially, who would want to live in a world where everyone thought like an engineer?

Whatever we do or say, our educational systems and environments foster and perpetuate specialties. They're designed to. From nature's viewpoint, there's no nanobioscience, no organometallic chemistry, no applied superconductivity, no condensed-matter physics; nor is there a thing called engineering. Such distinctions exist only in our minds, and too often they are powerful enough to guide or misguide our pro-

fessions. Cocooned in the belligerent provincialism of specialties and subspecialties, we ignore broader aspects of society.

Everyone is an engineer at some point in the way we design
our destinies. That's why it's the responsibility of not only
engineers, but just about everyone, to shape the future course
of engineering, which is entering an era of new eclecticism.
With a shared vision we can create better solution spaces,
convert random motions into progress, and improve socie-
tal muscle strength to address the complexities of today and
tomorrow.

4 Hitchcock was a backward thinker. His final
product was preordained but flexible. He valued
implementation over improvisation. "I make a film entirely on
paper. Not 'write it' but 'make it' on paper," he told a magazine
once, sounding much like an engineer working on a blueprint.
Hitchcock "came into a picture better prepared than almost
anybody I've ever seen," Jimmy Stewart said in an interview
for a French TV show. "He'd work on a script for five to six
months with the writer and become absolutely glib and absolutely know every scene, every word of the script."

With *The Birds*, as the script went, aggressive birds assaulted
the small town of Bodega Bay in northern California for reasons unknown. The main actors in this movie were the birds
themselves. The real birds were a raucous and expensive
bunch. They required a skilled trainer. Weekly food expenses

were about a thousand dollars. The birds ate about a hundred pounds of seeds each week, and twice that amount of anchovies, shrimp, and ground meat. The fake ravens were papier-mâché.

The technical issues with the movie were "prodigious," Hitchcock said. "I mean films like *Ben Hur* or *Cleopatra* are child's play compared to this. After all we had to train birds for every shot practically." Hitchcock's engineering background must have come in handy when he was applying the aerodynamic principles of gliders to simulate fake bird movements. Special wire works, miniatures, and gears were used to arrive at authentic-looking shots of the feathered actors—a Hollywood version of robotics before computers.

Hitchcock then faced the challenge of blending the separate foreground and background images of humans and birds. After exploring a number of possibilities—cobalt blue, sodium vapor, and infrared capture technologies—he eventually contracted with Walt Disney to use a photochemical yellow process in Technicolor. The outputs were splendid matte shots that could visually mass-produce the wing actions of both the real and fake birds with great synchronicity. These "special effects" would influence future smash hits like *Jaws* and *Jurassic Park*.

Then there was the question of sound design: Hitchcock was particular about a "toneless composition" that could be applied to the esoteric background sounds like the muttering of the crow or crying of the raven. "In *The Birds* there's a feeling of reality. You're really there," notes Gary Rydstrom, a seven-time Academy Award winner. To enhance the power of context, Hitchcock used a trautonium, an electronic musical device that generates vibratos. The result was *musique concrète*, a collage of synthesized music that enhanced the enigmatic mood.

In one scene of *The Birds* that Rydstrom highlights, the birds are gathering outside the school on a jungle gym before they attack a group of children. "There's this really ominous visual of this gathering storm, this gathering threat of these black birds. Other directors would have had ominous music, or ominous sound effects even," Rydstrom said, but "Hitchcock has singing children so that you have this contrast between the innocence of the victims that are going to be and the visual of the threat that's gathering onscreen. So he's telling you two different parts to the story, and it's much more effective." Like any creative engineer, Hitchcock not only understood the power of that context, but also combined his tools to optimize an effective outcome.

For *The Birds*, Hitchcock laid out detailed specifications for sound design. He stylized real sounds to "extract a little more drama out of ordinary sounds." The electronic sounds gave him "an extra use of dramatic device." The trautonium enabled him to experiment with contrasting extremes. In certain scenes, the electronic silence was profoundly effective. "Of course, I'm going to take the dramatic license of not having the birds scream at all," Hitchcock said in an interview with the French director François Truffaut. "I'm going to play the sound as though the birds are saying to this person,

'Now we've got you where we want you and here we come. We don't have to scream with triumph. We don't have to scream with anger. This is going to be a silent murder.'" This was not merely an effective use of technology, but an effective modulation of social psychology. In bridging that connection, Hitchcock was an authentic systems engineer.

Hitchcock never won an Oscar as a film director. His talent perhaps was never fully appreciated during his lifetime. "You sort of get a feeling that Hitch's life is as well planned as his movies," Stewart recounted in 1979 when the American Film Institute honored the director with a Lifetime Achievement Award. "Everything is laid out in advance right down to the last details. And like most people who began in silent films, it's the action that counts more than the words." And naturally, Hitchcock was a man of protocols. "I usually wear a blue suit, white shirt, black socks, no jewelry of any kind, no ornamentation, not even a wrist watch," he once said. "I think it relates to one's tidiness of mind. I have a very tidy mind."

In his fifty-odd movies, engineering and technology drove his stories forward, not the other way around. "Some directors film slices of life," Hitchcock observed. "I film slices of cake." By staying faithful to the story, he knew that he would maintain a deep bond with his audiences. If that link was weak, the results were nothing but gimmicks.

"You see, I like you. I want you to be happy. What more can I say?"

SOURCES AND RESOURCES

INTERVIEWS

This work benefited significantly from the insights and perspectives of Charlton Adams, G. D. Agarwal, José Andrés, Serge Appel, Norman Augustine, Celeste Baine, Simon Baron-Cohen, Craig Barrett, Harish Bhat, Vinton Cerf, Shu Chien, Francis Ching, Ralph Cicerone, Wayne Clough, David Collins, Robert Cook, Martin Cooper, Claire Curtis-Thomas, Ruth David, Tony DeRose, Gordon England, Robert Fein, Harvey Fineberg, Ralph Gomory, Ignacio Grossmann, Arlene Harris, Hank Hatch, Jim Hinchman, Chad Holliday, Ted Kaufman, David Koon, Robert Langer, George Laurer, Michael Lee, Steven Lien, Veer Bhadra Mishra, Vishwambar Nath Mishra, C. Dan Mote Jr., N. R. Narayana Murthy, Craig Newmark, José-Luis Novo, Vimla Patel, Atul Pawar, Charles Phelps, Jim Plummer, Nancy Pope, Bhaskar Ramamurthi, Steven Sasson, Piers Shepperd, Kenneth Shine, Barry Shoop, Ted Shortliffe, Robert Skinner Jr., Daniel Sniezek, Alfred Spector,

M. S. Swaminathan, Margaret Szymanski, Paul Tonko, Sophie Vandebroek, Charles Vest, John Viera, and George Whitesides.

NOTES

Prologue: Invisible Bridges

The Rosie Ruiz scandal was widely covered in the popular press and media. I culled information from several features published around the time of the event—and some follow-up articles published years later—in the *New York Times* (by Neil Amdur: October 25, 1976; October 23, 1983; by David Picker: November 7, 2005), in the *Washington Post* (by Jane Leavy: April 22, 23, and 27, 1980; by Lee Lescaze: April 25, 1980), in *Running Times* (by Ed Ayres: July 1980), in the *Christian Science Monitor* (by Greg Lamb: April 23, 1980), and in the *Evening Standard London* (by Adrian Warner: April 12, 2005). The interview between Ruiz and the reporter, as well as the eyewitness's quote, were transcribed from a 1980 news broadcast accessed on YouTube ("The Boston Marathon Cheater").

David Collins's 1994 book *Using Bar Code: Why It's Taking Over* (published by the Data Capture Institute) is an excellent source of information that combines his technical and social perspectives on a broad portfolio of bar code technologies that he and others have helped develop over time.

George Laurer's 2007 autobiography, *Engineering Was Fun* (published by Lulu.com), describes the development of the Universal Product Code. Quotes about the "golden chicken" (or "platinum pork") problem are from page 121. Information on the committee of grocery industry executives is from page 147. A technical paper coauthored by Laurer and David Savir—entitled "The Characteristics and Decodability of the Universal Product Code" (*IBM Systems Journal* 14, no. 1 [1975]: 16–34)—was instructive.

One of the first articles describing the impact of the UPC—"A Standard Labeling Code for Food"—was published in the April 7, 1973, issue of *Business Week* (pages 71–73). The 1999 PricewaterhouseCoopers report "17 Billion Reasons to Say Thanks: The 25th Anniversary of the U.P.C. and Its Impact on the Grocery Industry," by Vineet Garg and colleagues, served as a good reference. Economist Emek Basker's 2012 working paper for the National Bureau of Economic Research ("Raising the Barcode Scanner: Technology and Productivity in the Retail Sector") provided a good technical overview of how bar code scanners help increase productivity in the retail sector. The May 1974 *Harvard Business Review* article "The Grocery Industry in the USA—Choice of a Universal Product Code" was a helpful primer. The Retail Identification History Museum (ID History Museum) also contains a wealth of information regarding the UPC and related technologies. The 2011 book *Making the World Work Better: The Ideas That Shaped a Century and a Company*, by Kevin Maney, Steve Hamm, and Jeffrey O'Brien (IBM Press)—on IBM and its innovators—was a useful guide overall.

John Seabrook's quote is from the chapter "The Tower Builder" in his 2008 book *Flash of Genius* (St. Martin's Press), page 247.

One: Mixing and Matching

Historian Ken Alder's magnificent 1997 book *Engineering the Revolution: Arms & Enlightenment in France, 1763–1815* (University of Chicago Press) was the principal reference. Alder's 1991 dissertation, "Forging the New Order: French Mass Production and the Language of the Machine Age, 1763–1815" (Harvard

University); his 1997 paper "Innovation and Amnesia: Engineering Rationality and the Fate of Interchangeable Parts Manufacturing in France" (*Technology and Culture* 38, no. 2: 273–311); and Howard Rosen's 1981 dissertation, "The System Gribeauval: A Study of Technological Development and Institutional Change in Eighteenth Century France" (University of Chicago) provided additional historical background.

On the design of the naval carriage unit, see page 20 in Rosen's dissertation. The term "authentic military hero" is from page 22. The term "enlightenment engineering" is from the second chapter of Alder's book, the term "descriptionism" is quoted on page 71, and "mathematical gymnasium" is from page 90. "An enlightened man without passion . . ." is quoted from pages 38–39 of Alder's book—whose original source is "Rapport au Ministère, 3 March 1762" in Eugene Hennebert's *Gribeauval, lt-general de armees du roy* (Berger-Levrault, 1896), page 36.

The Louis XIV quote is from page 45 of John Lynn's chapter "Forging the Western Army in Seventeenth-Century France" in the 2001 book *The Dynamics of Military Revolution, 1300–2050*, edited by M. Knox and W. Murray (Cambridge University Press). The information about the number of horses and men required to drag a 34-pounder gun is also from this chapter (page 40). Winston Churchill's "first world war" quote—originally from his book *A History of the English-Speaking Peoples*—comes from the introduction to the 2014 book *The Culture of the Seven Years' War: Empire, Identity, and the Arts in the Eighteenth-Century Atlantic World*, edited by Shaun Regan and Frans De Bruyn (University of Toronto Press), page 3.

The cannon weight numbers are from the Xenophon Group—an association of military historians—based on research by historian Robert Selig accessed online on a web page headlined

"Statistical Overview of Artillery at the Siege of Yorktown (1781)."
The "Ordonnance Royale du 7 October 1732," reprinted in
Picard's *Artillerie française* (pages 55–56), is quoted on page 75 of
Alder's dissertation; and the quote "a system of control: rationality
made to serve despotism" comes from page 77. Gribeauval's background is drawn from Alder's book *Engineering the Revolution*, as
is the "Star Wars" quote (page 23).

Gribeauval's early years and technical contributions are also
recounted in Stephen Summerfield's December 2010 papers in
the *Smoothbore Ordnance Journal*: "Gribeauval's Early Work"
(pages 9–17); "Gribeauval in France before the Seven Years War
(1715–56)" (pages 18–23). More on the siege of Schweidnitz and
Gribeauval's translated text on Austrian army engineers can be
found in Summerfield's article "Gribeauval in Austrian Service
(1758–62)," published in the same issue of the *Smoothbore Ordnance Journal* (pages 24–35), in which he quotes (on page 26)
from Christopher Duffy's 2000 book *Instrument of War* (volume
1 of *The Austrian Army in the Seven Years War*; Emperor's Press).
Also helpful was a paper by Digby Smith: "Gribeauval Report
on the Austrian Artillery Dated 3 March 1762 (Translated from
Hennebert, Revue d'Artillerie, 1896, French)" (*Smoothbore Ordnance Journal*, December 2010, 63–66).

Supporting references included the following: Kevin Kiley,
"That Devil Gribeauval" (Napoleon Series Archive 2010, http://
www.napoleon-series.org); Bruce McConachy, "The Roots of
Artillery Doctrine: Napoleonic Artillery Tactics Reconsidered"
(*Journal of Military History* 65, no. 3 [2001]: 617–40); Ken Mac-
Lennan, "Liechtenstein and Gribeauval: 'Artillery Revolution' in
Political and Cultural Context" (*War in History* 10, no. 3 [2003]:
249–64); Brett Steele, "The Ballistics Revolution: Military and
Scientific Change from Robins to Napoleon" (PhD dissertation;

University of Minnesota, 1994); and Charles Gillispie and Ken Alder, "Exchange: Engineering the Revolution" (*Technology and Culture* 39, no. 4 [1998]: 733–54).

❖ ❖ ❖ ❖

Thomas Kuhn's classic book *The Structure of Scientific Revolutions* (University of Chicago Press, 3rd edition, 1996) and Freeman Dyson's *The Sun, the Genome, and the Internet: Tools of Scientific Revolutions* (Oxford University Press, 1999) provide stimulating yet somewhat contrasting perspectives on how science as an enterprise makes progress. Dyson's December 2012 article "Is Science Mostly Driven by Ideas or by Tools?" (*Science* 338, no. 6113: 1426–27) is also a useful read; it summarizes his view on how new tools—that is, engineering—help inform and underpin scientific progress. In 1939, MIT's Dugald Jackson went on to call engineering a "frontier influence" on modern civilization and social relations in a well-argued paper titled "Engineering's Part in the Development of Civilization" (*Science* 89, no. 2307: 231–37).

Tom Peters's quote is from "How Creative Engineers Think" (*Civil Engineering* 68, no. 3 [1998]: 48–51). Dan Mote's quote is from his presidential address entitled "Understanding Engineering," delivered at the fiftieth-anniversary celebration and 2014 annual meeting of the National Academy of Engineering, September 28, 2014. Other useful articles on engineering epistemology include Joseph Pitt's "What Engineers Know" (*Techné* 5, no. 3 [Spring 2001], available online); and Steven Goldman's "The Social Captivity of Engineering," in *Critical Perspectives on Nonacademic Science and Engineering*, edited by P. Durbin (Research in Technology Studies 4; Lehigh University Press, 1991), pages 121–45.

Henry Petroski's *The Essential Engineer: Why Science Alone*

Will Not Solve Our Global Problems (Vintage, 2011) also gives examples of how science sometimes gets in the way of engineering creations. For lateral reading on this topic, see also Edwin Layton's chapter "A Historical Definition of Engineering," also in Durbin's *Critical Perspectives* (pages 60–79). Other useful articles include a classic piece by W. F. Durand: "The Engineer and Civilization" (*Science* 42, no. 1615 [1925]: 525–33); and a more recent one by Zachary Pirtle entitled "How Models of Engineering Tell the Truth," in *Philosophy and Engineering*, edited by I. van de Poel and D. E. Goldberg (Springer, 2010), pages 97–108. Michael Davis's 1996 article "Defining 'Engineer': How to Do It and Why It Matters" (*Journal of Engineering Education* 85, no. 2: 97–101), and Steven Vick's 2002 book *Degrees of Belief: Subjective Probability and Engineering Judgment* (American Society of Civil Engineers Press) served as a useful framework for my thoughts.

Stuart Firestein and Andrew Wiles's quotes are from Firestein's 2012 book *Ignorance: How It Drives Science* (Oxford University Press), pages 2–7. On stepwise refinements, see Niklaus Wirth's classic paper "Program Development by Stepwise Refinement" (*Communications of the ACM*, April 1971, 221–27). Cognitive psychologists liken top-down and bottom-up design approaches to "big chunk" and "small chunk" cognitions involved in information processing. The term "hardware of culture" comes from Roger Burlingame's 1959 article by that title published in *Technology and Culture* (vol. 1, no. 1: 11–19).

The figure of less than 4 percent of (scientists and) engineers disproportionately creating jobs for the remainder is from the National Academies' 2010 report *Rising above the Gathering Storm, Revisited: Rapidly Approaching Category 5* (page 3). A related source is the National Science Board's *Science and Engineering Indicators 2014*. See also Nobel laureate Robert Solow's influential 1957 paper "Technical Change and the Aggregate

Production Function" (*Review of Economics and Statistics* 39, no. 3: 312–20), which showed how technological innovation alone is known to contribute more than 50 percent of economic growth. Robert Ayres's 1988 paper "Technology: The Wealth of Nations" (*Technological Forecasting and Social Change* 33: 189–201) is a useful read.

* * * *

Eugene Ferguson explains visual thinking in his 1992 book *Engineering and the Mind's Eye* (MIT Press). For more on structured thinking—including concepts such as "functional binding"—see Wayne Stevens, Glenford Myers, and Larry Constantine's 1974 paper "Structured Design" (*IBM Systems Journal* 13, no. 2: 115–39) and Barry Richmond's 1993 paper "Systems Thinking: Critical Thinking Skills for the 1990s and Beyond" (*Systems Dynamics Review* 9, no. 2: 113–33).

For systems thinking in general, I recommend, as examples, Peter Senge's revised and updated 2006 book *The Fifth Discipline: The Art & Practice of the Learning Organization* (Currency Doubleday); Peter Checkland's 1999 book *Systems Thinking, Systems Practice* (Wiley); Jay Forrester's books, including *World Dynamics* (Pegasus Communications, 1971), *Urban Dynamics* (MIT Press, 1969), and *Industrial Dynamics* (MIT Press, 1961); and Ludwig von Bertalanffy's classic 1969 work *General System Theory: Foundations, Development, Applications* (George Braziller).

The quote by Olivier de Weck and colleagues is from their 2012 book *Engineering Systems: Meeting Human Needs in a Complex Technological World* (MIT Press), page 34. For more on George Heilmeier's catechism, see his piece "Some Reflections on Innovation and Invention" (*Bridge* 22, no. 4 [1992]: 12–16). The checklist template comes from Joshua Shapiro's June 1994 profile of Heilmeier in *IEEE Spectrum* (page 58). A related article is by Chris Brantley: "The Heilmeier Catechism" (*IEEE-USA Today's Engineer*, February 2012).

Two: Optimizing

The Stockholm data are from a 2008 National Academies' Transportation Research Board conference proceedings—*U.S. and International Approaches to Performance Measurement for Transportation Systems*—from a section titled "Stockholm Congestion Charging Program: A Performance View," by Naveen Lamba (pages 84–85).

Lamba's quotes are from his article "Traffic and How to Avoid Future 'Carmageddons'" (*Fox Business*, July 15, 2011) and from a *Fast Company* piece listing 50 "Most Innovative Companies" that ranked IBM at 19 (February 10, 2009). Also insightful was an article by Lamba published in the *Tacoma News Tribune*, August 9, 2011: "New Technology Offers Solutions for Traffic Congestion." The *2012 Urban Mobility Report*, authored by David Schrank, Bill Eisele, and Tim Lomax, was produced by the Texas A&M Transportation Institute. The data are taken from page 5. U.K. statistics were taken from a TUC analysis, published November 14, 2015, of Labour Force Survey data for autumn quarters in 2008 and 2013.

Swedish transportation research professor Jonas Eliasson and his colleagues have published extensively about the effects of congestion pricing on public behavior in Stockholm. Sample works include Maria Börjesson et al.'s 2012 paper "The Stockholm Congestion Charges—Five Years On" (*Transport Policy* 20: 1–12) and Jonas Eliasson's 2009 paper "A Cost-Benefit Analysis of the Stockholm Congestion Charging System" (*Transportation Research Part A: Policy and Practice* 43, no. 4: 468–80).

The 2011 IBM report "Smarter Cities Series: Understanding the IBM Approach to Traffic Management," prepared by Stefen Schaefer et al.; and a 2013 paper by Akshay Vij and Joan Walker—"You Can Lead Travelers to the Bus Stop, but You Can't Make Them Ride" (*Transportation Research Board 92nd Annual Meeting Compendium of Papers*, National Academies Press)—were useful primers. Tom Vanderbilt's 2009 book *Traffic* (Vintage Books) effectively covers the psychology of—as the book's subtitle goes—*Why We Drive the Way We Do (and What It Says about Us)*.

* * * *

Statisticians George Box and Norman Draper famously declared that "all models are wrong, but some are useful" in their 1987 book *Empirical Model-Building and Response Surfaces* (John Wiley & Sons). The paper from which Joshua Epstein is quoted is "Why Model?" (*Journal of Artificial Societies and Social Simulation* 11, no. 4 [2008]: 12).

The "super-horse" story quoted from John Kuprenas and Matthew Frederick's 2013 book *101 Things I Learned in Engineering School* (Grand Central Publishing, page 75) was derived from "The Possibility of Life in Other Worlds," by Sir Robert Ball (*Scientific American Supplement*, no. 992 [January 5, 1895]: 15859–61). The 2001 book *Rational Analysis for a Problematic World Revisited*, by Jonathan Rosenhead and John Mingers (Wiley), is a good reference to learn more about such dichotomies as practical versus technical problems.

* * * *

Nancy Pope's quote "even if you're really good at it . . ." is from her July 1, 2013, interview "50 Years Ago, ZIP Codes Revolutionized Mail Service" on NPR's *All Things Considered*. A one-stop resource on the history and implementation of ZIP codes is the Smithsonian National Postal Museum's website (http://www .postalmuseum.si.edu). David Henkin's book *The Postal Age: The Emergence of Modern Communications in Nineteenth-Century*

America (University of Chicago Press, 2006) is one of the many terrific references and resources available for those interested in the history of postal communications.

❖   ❖   ❖   ❖

Quotes about Google Maps are from researchers Dragomir Anguelov et al., "Google Street View: Capturing the World at Street Level" (*Computer*, June 2010, 32–37). Also useful were Evan Ratliff's July 2007 article "The Whole Earth Cataloged: How Google Maps Is Changing the Way We See the World" in *Wired* and a July 2012 paper on "computational geography" by Carl Doersch et al.: "What Makes Paris Look Like Paris?" (*ACM Transactions on Graphics [SIGGRAPH 2012 Conference Proceedings]* 31, no. 4).

Randall Stephenson's quote is from Valentin Schmid's 2012 article "AT&T CEO Discusses Future of Mobile" (*Epoch Times*, November 22–28, A1). The Boeing 10-terabytes factoid is from a National Instruments report entitled "Automated Test Outlook 2013: A Comprehensive View of Key Technologies and Methodologies Impacting the Test and Measurement Industry" (page 10).

❖   ❖   ❖   ❖

Economist Gregory Mankiw's quote is from his 2006 paper "The Macroeconomist as Scientist and Engineer" (*Journal of Economic Perspectives* 20, no. 4: 29–46), which also cites John Maynard Keynes's quote on dentistry. "While the early macroeconomists were engineers trying to solve practical problems, the macroeconomists of the past several decades have been more interested in developing analytic tools and establishing theoretical principles. These tools and principles, however, have been slow to find their way into applications," Mankiw writes in this paper. "The tension between these two visions, while not always civil, may have been productive, for competition is as important to intellectual advance as it is to market outcomes." Another useful read is John Sutton's 2002 book entitled *Marshall's Tendencies: What Can Economists Know?* (MIT Press).

Eric Maskin's Nobel lecture is "Mechanism Design: How to Implement Social Goals" (December 8, 2007). See also articles by Alvin Roth—an operations researcher turned Nobel Prize–winning economist—including "The Economist as Engineer: Game Theory, Experimentation, and Computation as Tools for Design Economics" (*Econometrica* 70, no. 4 [2002]: 1341–78).

Alain Beltran's quote is from his 1993 book chapter "Competitiveness and Electricity: Electricité de France since 1946," in *Technological Competitiveness: Contemporary and Historical Perspectives on Electrical, Electronics, and Computer Industries*, edited by W. Aspray (IEEE Press), page 318, which also offers a great picture of the intersection of economic and engineering thinking.

A related paper, written by William Hausman and John Neufeld, is "Engineers and Economists: Historical Perspectives on the Pricing of Electricity" (*Technology and Culture* 30, no. 1 [1989]: 83–104). The concept of Boiteux's optimization is also closely related to what economists call "Ramsey pricing," after the brilliant mathematician Frank Ramsey, who died at the age of twenty-six. His works have provided a superb foundation for the modern scientific field of decision analysis.

For additional insights on how economists think, model, and analyze the world around them, I suggest two recent books: *The Assumptions Economists Make*, by Jonathan Schlefer (Belknap Press of Harvard, 2012); and Mary Morgan's *The World in the Model: How Economists Work and Think* (Cambridge University Press, 2012).

On differential pricing in theme parks, see, for example, Walter Oi's "A Disneyland Dilemma: Two-Part Tariffs for a Mickey Mouse Monopoly" (*Quarterly Journal of Economics* 85, no. 1 [1971]: 77–96). For insights on how the engineering design of the power industry has, in a way, led to an economic theory of such markets, see Stanford economist Robert Wilson's 2002 paper "Architecture of Power Markets" (*Econometrica* 70, no. 4: 1299–1340).

Three: Enhancing Efficiency and Reliability

On Clarence Saunders, the principal reference was Mike Freeman's excellent biography *Clarence Saunders & the Founding of Piggly Wiggly: The Rise & Fall of a Memphis Maverick* (History Press, 2011). The following quotes are from this book: "cut out all the frills of merchandising" (page 33); "every forty-eight seconds . . ." (page 34); "He liked to preach . . ." (page 35); "Every item is plainly marked . . ." (page 35); "In his hand, Piggly Wiggly was never . . ." (page 25). Saunders's U.S. patent number is 1242872. His other related patents are numbers 1357521, 1397824, 1407680, 1647889, and 1704061.

John Brooks's 1959 article "A Corner in Piggly Wiggly: Annals of Finance" (*New Yorker*, June 6, 128–50) and Henry Petroski's 2005 article "Shopping by Design" (*American Scientist*, November–December, 491–95) were useful references. Lauren Collins's quote is from her article "House Perfect: Is the IKEA Ethos Comfy or Creepy?" (*New Yorker*, October 3, 2011).

The term "category killers" is from Robert Spector's 2005 book *Category Killers: The Retail Revolution and Its Impact on Consumer Culture* (Harvard Business School Press). For more on IKEA history and philosophy, see Bertil Torekull's 1999 book *Leading by Design: The IKEA Story* (translated by Joan Tate;

Harper Business). Sam Walton's autobiography is *Sam Walton: Made in America* (Doubleday, 1992).

* * * *

Shepherd-Barron's quote about the chocolate bar dispenser idea is from Caroline Davies's obituary "Inventor of the Cash Machine Dies" (*Guardian*, May 19, 2010). The original source cited in this article is a 2007 BBC interview. Obituaries published in the *Los Angeles Times*, the *Guardian*, the *Windsor Star*, the *Herald Scotland*, and *BBC News* provided additional information on Shepherd-Barron.

Michael Lee's articles on the ATM Industry Association website were a tremendous source of information. Paul Volcker's comment on the ATM is from a December 14, 2009, *Wall Street Journal* article: "Think More Boldly; Interview at Future of Finance Initiative."

* * * *

On opportunistic assimilation, see the chapter by Colleen Seifert et al.—"Demystification of Cognitive Insight: Opportunistic Assimilation and the Prepared-Mind Hypothesis"—in *The Nature of Insight*, edited by R. Sternberg and J. Davidson (MIT Press, 1993), pages 65–124. On matrix thinking, see Tom Peters's "How Creative Engineers Think" (*Civil Engineering* 68, no. 3 [1998]: 48–51).

W. Bernard Carlson's quotes are from his chapter "Invention and Evolution: The Case of Edison's Sketches of the Telephone" in *Technological Innovation as an Evolutionary Process*, edited by J. Ziman (Cambridge University Press, 2000), pages 137–58.

Gary Bradshaw's article is "The Airplane and the Logic of Invention," in *Minnesota Studies in the Philosophy of Science*, vol. 15: *Cognitive Models of Science*, edited by R. Giere (University of Minnesota Press, 1992), pages 239–49. A related article, by Anthonie Meijers and Peter Kroes, is "Extending the Scope of the Theory of Knowledge," in *Norms in Technology*, edited by M. J. de Vries, S. O. Hansson, and A. W. M. Meijers (Springer, 2013), pages 15–34.

Tom Crouch's book on the Wright brothers is *The Bishop's Boys: A Life of Wilbur and Orville Wright* (W. W. Norton, 1989). Most of his quotes are from a November 11, 2003, PBS *Nova* article "The Unlikely Inventors." The quote "One literally has to 'see' the propeller . . ." is from Tom Crouch and Peter Jakab's 2003 book *The Wright Brothers and the Invention of the Aerial Age* (Smithsonian National Air and Space Museum, National Geographic), page 120.

W. Brian Arthur's 2005 paper is "The Logic of Invention" (Santa Fe Institute [SFI] Working Paper 2005-12-045:15). See also his 2009 book *The Nature of Technology: What It Is and How It Evolves* (Free Press).

* * * *

Eiji Toyoda's book is *Toyota: Fifty Years in Motion* (Kodansha International, 1987), and Taiichi Ohno's is *Toyota Seisan Hoshiki* (in Japanese; Diamond, 1978). Chapters 1 and 2 in David Magee's 2007 book *How Toyota Became #1: Leadership Lessons from the World's Greatest Car Company* (Portfolio) offered useful information. Eiji Toyoda's view on wringing water from a dry towel—as found in various articles and paraphrased here—appears to have been articulated in the context of cost-cutting efforts during the 1973 oil crisis, notes a February 26, 2010, Bloomberg piece by Alan Ohnsman and colleagues (available online). Escoffier's culinary contributions are drawn from Bee Wilson's 2012 book *Consider the Fork* (Basic Books), page 52.

Peter Senge's quote is from his 2006 book *The Fifth Disci-*

pline (Currency Doubleday), page 73. The Japanese sea model is described in Yuji Yamamoto and Monica Bellgran's paper "Fundamental Mindset That Drives Improvements towards Lean Production" (*Assembly Automation* 30, no. 2 [2010]: 124–30).

Information on reduced-weight airplane cutlery is from Mark Gerchik's 2013 book *Full Upright and Locked Position: Not-So-Comfortable Truths about Air Travel Today* (W. W. Norton), pages 247–49.

Taiichi Ohno's approach to root-cause analysis comes from his 1988 book *Toyota Production System: Beyond Large-Scale Production* (Productivity Press), pages 17–20. In nonscientific terms, the sort of questioning associated with root-cause analysis is not that different from the essence captured in the classic, rhythmic children's song "There's a Hole in My Bucket," about a circular conundrum involving two characters, Henry and Liza, sourced here from Wikipedia:

> *Henry has got a leaky bucket, and Liza tells him to repair it. But to fix the leaky bucket, he needs straw. To cut the straw, he needs a knife. To sharpen the knife, he needs to wet the sharpening stone. To wet the stone, he needs water. However, when Henry asks how to get the water, Liza's answer is "in a bucket." It is implied that only one bucket is available—the leaky one, which, if it could carry water, would not need repairing in the first place.*

Also useful as references on the Toyota Production System were James Womack and colleagues' 1991 book *The Machine That Changed the World: The Story of Lean Production* (Harper Perennial) and Allen Ward et al.'s 1995 article "The Second Toyota Paradox: How Delaying Decisions Can Make Better Cars Faster" (*MIT Sloan Management Review*, Spring, 43–61). One of the founding fathers of concurrent engineering—purely through the lens of efficiency—is Frederick Winslow Taylor. His 1911 book *The Principles of Scientific Management* (Harper & Brothers) is a good read. Martha Banta's 1995 book *Taylored Lives: Narrative Productions in the Age of Taylor, Veblen, and Ford* (University of Chicago Press) is also a good review of the various production philosophies related to concurrent engineering.

◆   ◆   ◆   ◆

For commentary on the RMS *Titanic*, see, for example, Henry Petroski's "100 Years after the Titanic, We're Still Not Unsink-able" (*Washington Post*, April 6, 2012). On aggressive and conservative trade-offs, Kevin Otto and Erik Antonsson's 1991 paper "Trade-Off Strategies in Engineering Design" (*Research in Engineering Design* 3, no. 2: 87–104) is a good reference.

Regarding social trust of engineering, perhaps one of the oldest rules is the Code of King Hammurabi, which—as quoted on page 41 of John Kuprenas and Matthew Frederick's 2013 book *101 Things I Learned in Engineering School* (Grand Central Publishing)—states:

> *If a builder has built a house for a man, and has not made it sound, and the house falls and causes the death of its owner, that builder shall be put to death. If it is the owner's son that is killed, the builder's son shall be put to death. If it is the slave of the owner that is killed, the builder shall give a slave to the owner of the house. If it ruins goods, the builder shall make compensation for all that has been ruined, and shall re-erect the house from his own means. If a builder builds a house, even though he has not yet completed it; if then the walls seem toppling, the builder must make the walls solid from his own means.*

Michio Kaku's concept of High Tech and High Touch comes from his 2011 book *Physics of the Future: How Science Will Shape Human Destiny and Our Daily Lives by the Year 2100* (Doubleday), page 15.

Four: Standardizing with Flexibility

Alexander Fleming's 1929 paper is "On the Antibacterial Action of Cultures of a Penicillium, with Special Reference to Their Use in the Isolation of *B. influenzae*" (*British Journal of Experimental Pathology* 10, no. 3: 226–36). Fleming's quotes come from his December 10, 1945, speech at the Nobel banquet and his Nobel lecture "Penicillin," December 11, 1945, in Stockholm—both accessible on the Nobel Prize website. The term "national hero" is from Richard Cavendish's article "Funeral of Sir Alexander Fleming" (*History Today* 55, no. 3 [March 2005]).

The history of penicillin mass production has been widely reported. See, for example, Maxwell Brockmann et al.'s 1970 book *The History of Penicillin Production* (American Institute of Chemical Engineers); Pfizer, Inc.'s 2008 booklet *Penicillin Production through Deep-Tank Fermentation* (National Historic Chemical Landmarks program of the American Chemical Society); Erik Lax's 2004 book *The Mold in Dr. Florey's Coat: The Story of the Penicillin Miracle* (Henry Holt); David Wilson's 1976 book *In Search of Penicillin* (Alfred A. Knopf); Joseph Lombardino's 2000 paper "A Brief History of Pfizer Central Research" (*Bulletin for the History of Chemistry* 25, no. 1: 10–15); and sections of two National Research Council reports—*Frontiers in Chemical Engineering: Research Needs and Opportunities* (1988) and *Separation & Purification: Critical Needs and Opportunities* (1987)—both from the National Academy Press.

* * *

Representative sources by Hugh De Haven that I consulted include "Mechanical Analysis of Survival in Falls from Heights of Fifty to One Hundred and Fifty Feet" (*Injury Prevention* 6 [2000]: 62–68), which originally appeared in *War Medicine* 2 (1942): 586–96; and *Development of Crash-Survival Design in Personal, Executive, and Agricultural Aircraft* (Crash Injury Research, Cornell University Medical College, 1953).

For commentaries on De Haven's research implications, see Amy Gangloff's paper "Safety in Accidents: Hugh DeHaven and the Development of Crash Injury Studies" (*Technology and Culture* 54, no. 1 [2013]: 40–61) and Carl Metzgar's "Writing Worth Reading: Mechanical Analysis of Survival in Falls from Heights of 50 to 150 Feet" (*Professional Safety*, June 2003, 55, 75). The alternative account of the 1916 crash story is that the other pilot managed to walk away from his wrecked aircraft while De Haven was left debilitated. I relied on Gangloff's documentation, which in turn relied on De Haven's own "good record keeping."

The quotes "the package should not open up . . . ," "from impact against the inside . . . ," "would not test . . . ," "fragile, valuable objects . . . ," "extreme forward movement . . . ," and "fully appreciate the fact that . . . ," are from De Haven's 1952 SAE Technical Paper 520016: "Accident Survival—Airline and Passenger Car," reprinted in William Haddon, Edward Suchman, and David Klein's 1964 *Accident Research: Methods and Approaches* (Harper & Row), pages 562–68. Howard Hasbrook's quote is from his 1956 paper "The Historical Development of the Crash-Impact Engineering Point of View" (*Clinical Orthopaedics* 8: 268–74), reprinted in Haddon et al.'s book, pages 547–54.

For more on the CREEP framework, see the 2002 book *Small Airplane Crashworthiness Design Guide*, edited by Todd Hurley and Jill Vandenburg (Simula Technologies), pages 1–3. On the invention of the three-point seat belt, see the U.S. patent issued to Roger Griswold and Hugh De Haven: "Combination Shoulder and Lap Safety Belts," no. 2710649, June 14, 1955. The description of the fighter pilots' shoulder harness as "uncomfortable and overly restricting" comes from this patent. A crosscutting 2012 volume edited by Sabine Roeser et al. (*Handbook of Risk Theory: Epistemology, Decision Theory, Ethics, and Social Implications of Risk*; Springer) and a 1958 review article by Charles Chayne ("Automotive Design Contributions to Highway Safety," *Annals of the American Academy of Political and Social Science* 320, no. 1: 73–83) offered valuable insights.

The U.S. Centers for Disease Control and Prevention articles on seat belts include "Achievements in Public Health, 1900–1999 Motor-Vehicle Safety: A 20th Century Public Health Achievement" (*Morbidity and Mortality Weekly Report*, May 14, 1999, 369–74); "Seat Belts Fact Sheet" at http://www.cdc.gov/motor vehiclesafety; and "Adult Seat Belt Use in the US," in *CDC Vital Signs* (National Center for Injury Prevention and Control, 2011).

* * * *

Information about Margaret Hutchinson was gathered primarily from reporting in the *Daily Boston Globe* (July 29, 1952; June 26, 1955; December 9, 1956), the *Christian Science Monitor* (June 7, 1950), and the *New York Times* (June 25, 1955).

The quote "Heating, cooling, washing . . ." is from James Sparkman's "Chemical Engineering Family: 'Home' Is Extensive Place for Woman Chemistry Sc.D." (*Christian Science Monitor*, June 7, 1950); "keeps us so busy we don't have . . ." comes from Nat Kline's "Two at Stone & Webster Fear No Male Competition" (*Daily Boston Globe*, July 29, 1952); "I actually received little . . ."

is from the article titled "Woman Engineer from Reading Wins Top Award" (*Daily Boston Globe*, June 26, 1955).

Pfizer executive John L. Smith's quote "The mold is as temperamental . . ." is from the American Chemical Society's "National Historic Chemical Landmarks" web page on the discovery and production ramp-up of penicillin. Penicillin production data are from A. N. Richards's 1964 article "Production of Penicillin in the United States (1941–1946)" (*Nature* 201: 441–45). The quote "far beyond the art of brewing" is from the *Hartford Courant*'s article "World's Largest Output of Antibiotics at Groton" (May 1, 1955).

◆　◆　◆　◆

On compound technologies, see Cambridge University industrial economist Antonio Andreoni's works, including his chapter "On Manufacturing Development under Resources Constraints" in *Resources, Production and Structural Dynamics*, edited by M. Baranzini, C. Rotondi, and R. Scazzieri (Cambridge University Press, 2015).

The term *interpretive consistency* was coined by Gregory Gargarian. See, for example, his 1996 piece "The Art of Design," in *Constructionism in Practice: Designing, Thinking, and Learning in a Digital World*, edited by Y. Kafai and M. Resnick (Lawrence Erlbaum), page 136.

The example of India's "local systems of measures" is from Lawrence Busch's 2011 book *Standards: Recipes for Reality* (MIT Press), page 61. For more on India's 1956 Standards of Weights and Measures Act, see the volume *Metric Change in India*, edited by Lal Verman and Jainath Kaul (Indian Standards Institution, 1970), page 306.

Donald Berwick's quote is from "How to Fix the System" (*Time*, April 24, 2006). On sirens, see Brian Hills's "Vision, Visibility, and Perception in Driving" (*Perception* 9, no. 2 [1980]: 183–216), as well as a review paper by Robert De Lorenzo and Mark Eilers: "Lights and Siren: A Review of Emergency Vehicle Warning Systems" (*Annals of Emergency Medicine* 20, no. 12 [1991]: 1331–35). On the building of complicated systems, I recommend the 2011 book by Olivier de Weck, Daniel Roos, and Christopher Magee titled *Engineering Systems: Meeting Human Needs in a Complex Technological World* (MIT Press); see, for example, page 32.

John Rae's quote is from his 1960 paper "The 'Know-How' Tradition: Technology in American History" (*Technology and Culture* 1, no. 2: 139–50). Useful companion articles included Carolyn Miller's 1998 piece "Learning from History: World War II and the Culture of High Technology" (*Journal of Business and Technical Communication* 12, no. 3: 288–315) and Thomas Hughes's 2004 book *American Genesis: A Century of Invention and Technological Enthusiasm, 1870–1970* (University of Chicago Press).

With Michael Vincent and Kenneth McLeod, I discuss the concept of transduction, among other biological metaphors (including transformation and fusion, which come up in the chapter on prototyping), in a chapter entitled "Evolutionary Processes as Conceptual Metaphor for Innovative Design Processes in Engineering," published in the 2007 book *Innovations 2007: World Innovations in Engineering Education and Research*, edited by W. Aung et al. (International Network for Engineering Education and Research Press), pages 441–52. In biological fusion the integrated technologies are not merely add-ons, but they become part of the system, resulting in emergent behavior. Ricard Solé and colleagues explore additional related ideas in their 2013 white paper "The Evolutionary Ecology of Technological Innovations" (Santa Fe Institute).

On the innovation of Harold Willis and Henry Ford—and several other engineering innovators—see the section on Henry

Ford in Harold Evans's 2006 *They Made America: From the Steam Engine to the Search Engine, Two Centuries of Innovators* (Back Bay Books). The quote is from page 303.

Recommended companion books include Douglas Brinkley's *Wheels for the World: Henry Ford, His Company, and a Century of Progress* (Penguin, 2004), Lindy Biggs's *The Rational Factory: Architecture, Technology and Work in America's Age of Mass Production* (Johns Hopkins University Press, 1996), and David Nye's *America's Assembly Line* (MIT Press, 2013).

Five: Solutions under Constraints

Varanasi is also known as Banaras or Kashi. Mark Twain's quote is from his 1897 book *Following the Equator: A Journey around the World* (American Publishing Company). Other prominent discussions on Varanasi include Richard Lannoy's 1999 book *Benares Seen from Within* (Callisto Books); E. B. Havell's 1905 volume *Benaras: The Sacred City* (Vishwavidyalaya Prakashan, reprinted 1990); Jonathan Parry's 1994 work *Death in Banaras* (Cambridge University Press); Stephen Alter's 2001 book *Sacred Waters: A Pilgrimage up the Ganges River to the Source of Hindu Culture* (Harcourt); and Mark Tully's chapter "Varanasi: The Unity of Opposites," in his 2007 book *India's Unending Journey* (Rider).

Harvard's Diana Eck calls Varanasi the "crossing place" in her notable 1998 book *Banaras: City of Light* (Columbia University Press). The quote "Death in Kashi [Varanasi] . . ." is from page 24. The description of Manikarnika Ghat as the "great cremation ground" is from page 32.

The fact that the water of the Ganges and dwelling in Varanasi are two of the most substantial things is paraphrased from the eighth-century Indian saint-philosopher Adi Sankara, noted in Veer Bhadra Mishra's 2005 paper "The Ganga at Varanasi and a Travail to Stop Her Abuse" (*Current Science* 89, no. 5: 755–63). Mishra's January 2013 monograph *Wastewater Management in Ganga Basin* (Kishor Vidya Bhavan) offers additional insights and ideas. Mishra's quote "not a difficult problem" is from his 2010 TEDxDelhi talk.

The snakes-and-ladders quote ("My campaign has been like . . .") is taken from the website of the Sankat Mochan Foundation. Alexander Stille's 1998 article "The Ganges' Next Life" (*New Yorker,* January 19) is a superb profile of Mishra. Also useful was an article written by the late Fran Peavey entitled "The Birth of Cleaning the Ganges Project," available at http://www.crabgrass.org.

Dean Young's lines are from his 2005 collection of poems entitled *Elegy on Toy Piano* (University of Pittsburgh Press). I originally spotted Young's words in Geoff Dyer's 2010 book *Jeff in Venice, Death in Varanasi* (Vintage), which contains the quote on Varanasi traffic (page 164).

The coliform levels are from "Creaking, Groaning: Infrastructure Is India's Biggest Handicap" (*Economist*, December 11, 2008). The World Health Organization provides relevant information on the "Global Epidemics and Impact of Cholera" page of its website.

Alan Turing's quote can be found in various forms online, but the original source appears to be "Epigram to Robin Gandy" (1954), reprinted in Andrew Hodges's book *Alan Turing: The Enigma* (Vintage, 1992), page 513, as noted by Wikipedia.

President Bill Clinton's remarks can be found at the American Presidency Project maintained by the University of California, Santa Barbara; see "Remarks to the Business Community in Hyderabad, March 24, 2000." William Oswald's papers offer background on the technical solutions that Mishra adapted and proposed for Varanasi: "Ponds in the Twenty-First Century" (*Water Science and Technology* 31, no. 12 [1995]: 1–8) and "Introduction to Advanced Wastewater Ponding System" (*Water Science and Technology* 24 [1991]: 1–7).

Mishra's quote "I pray that I should be able . . ." is from David Suzuki's documentary *The Sacred Balance* (Kensington Communications), in an excerpt accessed on YouTube. Also recommended is a 1998 documentary entitled *Holy Man and Mother Ganga* (41st Floor Films), written and produced by Patricia Chew and Claude Adams.

The woman in the hospice and Shiva spoke to me in Hindi, which I translated into English.

❖ ❖ ❖ ❖

The historical details—including the quote "made a feeble effort to rise . . ."—are from Sandford Fleming's 1877 report, which is cited and discussed in the book *Sir Sandford Fleming*, edited by Doris Unitt, Andrew Osler, and Edward McCoy (Clockhouse, 1968), pages 74–75.

The Ireland train station story is from Clark Blaise's terrific 2001 book *Time Lord: The Remarkable Canadian Who Missed His Train, and Changed the World* (Vintage Canada), pages 66, 75–76. Fleming's contemporary was Charles Dowd, and the quote comes from page 95 of Blaise's book. The 144-time-zones factoid and the term "bloodthirsty savage" are also from this book.

Some have attributed Greenwich Mean Time as a prime force leading to Britain's industrial development, putting it in the lead—perhaps nearly half a century ahead of others. But some parts of the world have resisted standardizing to Greenwich Mean Time, for various reasons. Arizona, for example, does not observe daylight savings time; and some countries, like India, are thirty minutes off. China, which has an impressive span that would ordinarily encompass five time zones, has only one time. A web search on these topics will yield more information.

Another important person in the history of time zones is meteorologist Cleveland Abbe. Interested readers may wish to read a brief biographical sketch, written by J. Humphreys in 1918, entitled "Biographical Memoir of Cleveland Abbe, 1838–1916" (National Academy of Sciences Biographical Memoirs 8, pages 469–508).

The term "culture's time" and the quote "There was no 'system' . . ." are from Ian Bartky's 1989 paper "The Adoption of Standard Time" (*Technology and Culture* 30, no. 1: 25–56). The railway miles expansion data from 1832 and 1880 are cited in this paper with a source directing to the *Encyclopaedia Britannica* entry published in 1943. Bartky has also written two excellent scholarly volumes published by Stanford University Press: *Selling the True Time: Nineteenth-Century Timekeeping in America* (2000) and *One Time Fits All: The Campaigns for Global Uniformity* (2007).

Other useful references include Lawrence Burpee's 1915 *Sandford Fleming, Empire Builder* (Humphrey Milford), Hugh Maclean's 1969 *Man of Steel: The Story of Sir Sandford Fleming* (Ryerson Press), and David Prerau's 2006 *Seize the Daylight: The Curious and Contentious Story of Daylight Saving Time* (Basic Books).

❖   ❖   ❖   ❖

Former Indian president A. P. J. Abdul Kalam's quotes are taken from his valedictory address at the Annual National Techno-Management Symposium, Jaipur, February 26, 2012. Full text of his speech is available on his website. The Samuel Johnson quote is from James Boswell's 1791 book *Life of Johnson*, abridged and edited with an introduction by Charles Grosvenor Osgood (Electronic Classics Series, recently issued by Pennsylvania State University), page 285.

Kevin Kelly's quote is from his 2010 book *What Technology Wants* (Viking Penguin), page 110. John Armitt's quote is from Nick Smith's July 2012 piece "Setting the Stage for the Greatest Show on Earth" (*IET Member News*, 10–13). More on constraint programming can be found in the 2006 *Handbook of Constraint Programming*, edited by Francesca Rossi, Peter van Beek, and Toby Walsh (Elsevier Science).

* * * *

Gordon Cook's quote on Bazalgette is from "Construction of London's Victorian Sewers: The Vital Role of Joseph Bazalgette" (*Postgraduate Medical Journal* 77 [2001]: 802–4). Stephen Halliday's 2009 book *The Great Stink of London: Sir Joseph Bazalgette and the Cleansing of the Victorian Metropolis* (History Press) is a comprehensive resource.

I added emphasis on "nothing happens" in the quote "There are television shows . . ." by Harvey Fineberg, which was published in "The Paradox of Disease Prevention" (*JAMA* 310, no. 1 [2013]: 85–90). See also his informative article "Public Health and Medicine: Where the Twain Shall Meet" (*American Journal of Preventive Medicine* 41, no. 4S3 [2011]: S149-51).

Six: Crossing Over and Adapting

David Koon gives a first-hand account of his daughter Jennifer's murder and his transition from engineering to politics in a chapter entitled "Politics and Legislation," in *Career Development in*

Bioengineering and Biotechnology, a book I coedited with Barbara Oakley and Luis Kun (Springer, 2008), pages 233–38. Also useful as references were articles and analyses by Tanya Fluette for the *Rochesterian*, by Lois Rumfelt for the *Lutheran*, by David Schneider for the *American Scientist*, and by others for the *Buffalo News*, *Wireless Review*, *Mobile Radio Technology*, and the *Rochester Democrat and Chronicle*. Koon's *Newsday* quote on death and dying ("Even her voice . . .") is from Bret Begun's article "Pain of Murder Victim's Parents" (August 16, 1997).

My description of the 2003 rowboat accident involving four teenagers is based on reporting for the *New York Times* by Al Baker, Sheila Dewan, Kevin Flynn, Robert McFadden, Lydia Polgreen, William Rashbaum, Marc Santora, Michael Wilson, and Robert Worth. The information about the teenagers buying cookies, sweets, and Frappuccino, for instance, is from Santora's piece "Facing Icy Waters and Grim Realities" (January 29, 2003).

Koon's quote "I put their deaths on our governor's shoulders" is mentioned in Glenn Bischoff's article "An Ounce of Prevention" (*Wireless Review*, May 2003, 32–33, 41). The hero quote is from Glenn Bischoff's "Wavelengths: Koon Truly Deserving of 'Hero' Award" (*Mobile Radio Technology Bulletin*, March 5, 2004). One of Koon's high-profile senate testimonies (made together with the then senator Hillary Clinton) on "E-911 Implementation" was with the Communications subcommittee of the Committee on Commerce, Science, and Transportation in 2003.

◆   ◆   ◆   ◆

On the "senate syndrome," see Steven Smith's 2010 "The Senate Syndrome" (Brookings Institution, Issues in Governance Studies, June). For an extended treatment on this subject, see Smith's 2014 book under the same title (University of Oklahoma Press).

The discussion of selection bias in politics is based on the article "There Was a Lawyer, an Engineer and a Politician . . . Why Do Professional Paths to the Top Vary So Much?" (*Economist*, April 16, 2009).

In place of *frame mismatch*, philosophers may use the more formal term *frame problem*, which, as noted in the *Stanford Encyclopedia of Philosophy*, "is the challenge of representing the effects of action in logic without having to represent explicitly a large number of intuitively obvious non-effects." Abraham Lincoln's view on public sentiment has been discussed in countless publications. My source is "The Ottawa Debate" in the 1991 edition of *The Complete Lincoln-Douglas Debates of 1858*, edited by P. Angle (University of Chicago Press; originally published in 1958), page 128.

Simon Baron-Cohen's quote is from page 103 of the 1997 paper "Is There a Link between Engineering and Autism?" (*Autism* 1, no. 1: 101–9), which he coauthored with several colleagues. For further reading, I suggest Baron-Cohen's more recent article titled "Autism and the Technical Mind" (*Scientific American*, November 2012, 72–75). Much research has been carried out—and recently popularized—on dichotomies in the human brain. In addition to Baron-Cohen's research, I would point readers to popular accounts by Nobel laureate Daniel Kahneman in *Thinking, Fast and Slow* (Farrar, Strauss and Giroux, 2011) and by Iain McGilchrist in *The Master and His Emissary: The Divided Brain and the Making of the Modern World* (Yale University Press, 2009).

The Dilbert quote is from James Braham's 1992 article "The Silence of the Nerds" (*Machine Design*, August 20, 75–80). Steve Wozniak's quote is taken from radio host and former engineer Ira Flatow's 2008 book *Present at the Future* (HarperCollins) in a chapter entitled "The Wizard of the Woz" (pages 259–60).

Astronaut Neil Armstrong's quote comes from his essay "The Engineered Century," published in the National Academy of Engineering periodical *The Bridge* (Spring 2000, 14–18).

Seven: Prototyping

Steve Sasson's quote "our plan was unrealistic . . ." is from his 2010 talk at the Chautauqua Institution. Eastman's motto on the camera being as "convenient as the pencil" is from Kodak's website. Elizabeth Brayer's *George Eastman: A Biography* (Johns Hopkins University Press, 1996) and Todd Gustavson's *Camera: A History of Photography from Daguerreotype to Digital* (Sterling Signature; reprint, 2012) are useful sources of historical information. The quote "basically told Sasson . . ." is by Timothy Lynch, Kodak's chief intellectual property officer, taken from Mark Harris's "Snapping Up Kodak" (*IEEE Spectrum*, February 2014, 30–35).

For an accessible history of mobile phones, see "1973–1983: Making History, Developing the Portable Cellular System" under "Cell Phone Development" on the history page of the Motorola Solutions website (http://www.motorolasolutions.com). Also useful was Motorola's Press Information piece entitled "The Cellular Telephone Concept—An Overview," September 10, 1984. The quote "it will be possible . . ." is from a Motorola press release issued in New York, April 3, 1973: "Motorola Demonstrates Portable Telephone to Be Available for Public Use by 1976." See also the following articles on Martin Cooper: "Father of the Cell Phone" (*Economist*, June 4, 2009); Tas Anjarwalla's "Father of the Cell Phone" (*CNN Tech*, July 9, 2010); Pagan Kennedy's "Who Made That Cellphone" (*New York Times*, March 15, 2013).

"We realize the impossible . . ." is from the website of Wonder Works, Piers Shepperd's company (http://www.wonder.co.uk). On

Einstellung, see, for example, a 2010 review by Merim Bilalić, Peter McLeod, and Fernand Gobet: "The Mechanism of the Einstellung (Set) Effect: A Pervasive Source of Cognitive Bias" (*Current Directions in Psychological Science* 19, no. 2: 111–15). John Lienhard's analysis of technological performance enhancements can be found in his 1979 paper "The Rate of Technological Improvement before and after the 1830s" (*Technology and Culture* 20, no. 3: 515–30). See also "Rates of Technological Improvement: Doubling in a Lifetime," episode 559 of Lienhard's radio program *The Engines of Our Ingenuity* (Houston Public Media), and his 1985 paper "Some Ideas About Growth and Quality in Technology" (*Technological Forecasting and Social Change* 27: 265–81).

A concept close to prototyping is reverse engineering. The U.S. Supreme Court has recognized reverse engineering as an "essential part of innovation" that could lead to important technological advances. More on this topic can be found in Pamela Samuelson and Suzanne Scotchmer's 2002 review entitled "The Law and Economics of Reverse Engineering" (*Yale Law Journal* 111: 1575–1662).

Eight: Learning from Others

Victor Mills's "mess" quote is from Jeff Harrington's 1997 article "Disposable Diaper Inventor Dies" (*Cincinnati Enquirer,* November 7). Another illustrative article on Mills is "Victor Mills Is Dead at 100; Father of Disposable Diapers," by Andrew Revkin (*New York Times,* November 7, 1997). The information about Betsy Wetsy comes from an article by Claudia Flavell-While: "A 'Pampered' Career" (*Chemical Engineer Today,* May 2011, 52–53). Malcolm Gladwell, in "Smaller: The Disposable Diaper and the Meaning of Progress" (*New Yorker,* November 26, 2001), discusses the lesser-known development of Huggies, the rival of Pampers.

✦   ✦   ✦   ✦

The ketchup case study is from the chapter "Investigating and Researching for Design Development" in the book *Engineering Design: An Introduction*, by John Karsnitz, Stephen O'Brien, and John Hutchinson (Delmar Cengage Learning, 2nd edition, 2013), pages 186–87. Related references include the article "Heinz, Hunt's Turn Ketchup Upside-Down" (*Packaging World*, June 30, 2002) and Arnie Orloski's "Heinz 'Caps' Squeeze Ketchup" (*Packaging World*, April 30, 2000). The NASCAR pit stop quote is taken from an article written by journalist Frank Greve: "Top-Down Approach Rekindles Our Love Affair with Ketchup" (*Seattle Times*, June 27, 2007).

✦   ✦   ✦   ✦

For an account of the historically low enrollment of women in engineering, see, for example, Amy Sue Bix's 2014 book *Girls Coming to Tech!: A History of American Engineering Education for Women* (MIT Press).

Lucy Suchman's views on user-centered design come from her 2006 book *Human-Machine Reconfigurations: Plans and Situated Actions* (Cambridge University Press, 2nd edition), pages 9, 23. See also a related chapter, "Work Practice and Technology: A Retrospective," which Suchman wrote in *Making Work Visible: Ethnographically Grounded Case Studies of Work Practice*, edited by M. Szymanski and J. Whelan (Cambridge University Press, 2011), pages 21–33. Another broad analysis of this subject is presented by Bruno Latour in his 1988 book *Science in Action: How to Follow Scientists and Engineers through Society* (Harvard University Press). A good account of Xerox culture is Michael Hiltzik's *Dealers of Lightning: Xerox PARC and the Dawn of the Computer Age* (Harper Business, 2000).

✦   ✦   ✦   ✦

Sony chairman Akio Morita's quotes are from his 1986 memoir *Made in Japan: Akio Morita and Sony* (Dutton), pages 64, 65, 264. The phrase "preserved women" is from social historian Shelley Nickles's 2002 paper "'Preserving Women': Refrigerator Design as Social Process in the 1930s" (*Technology and Culture* 43, no. 4: 693–727). For a discussion of other cultural effects of refrigerators, see Bee Wilson's 2012 book *Consider the Fork* (Basic Books).

The article entitled "The House of Quality" is by John Hauser and Don Clausing (*Harvard Business Review*, May–June 1988). The *Los Angeles Sentinel* article (Ann Job, Behind the Wheel, "Toyota Avalon," April 13, 2000) contains the phrase "stone-pecking noise." Michael Kennedy's quote is from Leland Teschler's article "How to Develop Products Like Toyota" (*Machine Design*, October 9, 2008, 58–64).

For more on the opening of Japan in the mid-1800s, I recommend the U.S. Navy Museum's website (one sample article is "Commodore Perry and the Opening of Japan," available at http://www.history.navy.mil). The idea of engineering needing a "black ship effect" occurred to me when I was reading a thoughtful chapter by Koji Kishimoto—"Fujitsu Learned Ethnography from PARC: Establishing the Social Science Center"—in Szymanski and Whelan's *Making Work Visible* (pages 327–35). Kishimoto likens Fujitsu to Japan's process of liberation, and I adapted and applied that idea to advance my case of how engineering could benefit from its own version of creative liberation. Kishimoto also summarizes three ethnographic processes that I believe can be applied for innovative outcomes in engineering or any other creative process: field observation, reflection, and codesign.

✦   ✦   ✦   ✦

Francisco Aguilera's paper is titled "Is Anthropology Good for the Company?" (*American Anthropologist* 98, no. 4 [1996]: 735–42). See also Erik Styhr Petersen, James Nyce, and Margareta Lützhöft's 2011 paper "Ethnography Re-engineered: The Two Tribes Problem" (*Theoretical Issues in Ergonomics Science* 12, no. 6: 496–509). Diana Forsythe's views are from her 1999 paper "'It's Just a Matter of Common Sense': Ethnography as Invisible Work" (*Computer Supported Cooperative Work* 8: 127–45). For further inquiry, the following well-regarded works should be helpful references: Hortense Powdermaker's *Stranger and Friend: The Way of an Anthropologist* (W. W. Norton, 1967); and Clifford Geertz's *The Interpretation of Cultures* (1977) and *Local Knowledge: Further Essays in Interpretive Anthropology* (1983), both published by Basic Books.

Mamie Warrick's quote is from Jeffrey Liker and Michael Hoseus's 2008 book *Toyota Culture: The Heart and Soul of the Toyota Way* (McGraw-Hill), page 326. In his article "Drink Me: How Americans Came to Have Cup Holders in Their Cars" (*Slate*, March 15, 2004), Henry Petroski discusses how nonessential features subsequently become indispensable in the interior design of automobiles. For research on "reductive bias" and oversimplification in engineering design, see the 2004 paper by Paul Feltovich et al. entitled "Keeping It Too Simple: How the Reductive Tendency Affects Cognitive Engineering" (*IEEE Intelligent Systems* 19, no. 3 [May/June]: 90–94).

*   *   *   *

In his 1999 book *Beyond Engineering: How Society Shapes Technology* (Oxford University Press), Robert Pool eloquently analyzes how culture shapes engineering outcomes. Marketing scholar John Sherry's 1986 paper "The Cultural Perspective in Consumer

Research" (*Advances in Consumer Research* 13: 573–75) is a good companion read, as is Jodi Forlizzi's 2008 piece "The Product Ecology: Understanding Social Product Use and Supporting Design Culture" (*International Journal of Design* 2, no. 1: 11–20). For the role of listeners in influencing musical evolution, see, for example, the paper "Evolution of Music by Public Choice," by Robert MacCallum et al., and the related commentary by Christoph Adami ("Adaptive Walks on the Fitness Landscape of Music"), both published in the *Proceedings of the National Academy of Sciences* (109, no. 30 [2012]).

On the concept of preference paradox, see Nobel Prize–winning economist Kenneth Arrow's 1950 paper "A Difficulty in the Concept of Social Welfare" (*Journal of Political Economy* 58, no. 4: 328–46), as well as his 1963 book *Social Choice and Individual Values* (John Wiley & Sons, 2nd edition).

There appears to be no documented evidence that Henry Ford actually said his customers wanted faster horses, but the quote is often attributed to him. See, for example, an interesting piece by Patrick Vlaskovits: "Henry Ford, Innovation, and That 'Faster Horse' Quote" (*Harvard Business Review* blog, August 29, 2011). Steve Jobs's quotes come from the article "Apple's One-Dollar-a-Year Man" (*Fortune*, January 24, 2000). Marissa Mayer's quote is from her conversation with Erik Schatzker during the session "An Insight, an Idea with Marissa Mayer" at the 2013 World Economic Forum Annual Meeting, Davos, Switzerland.

❖   ❖   ❖   ❖

Anthropologist Stephen Lansing's quotes are from his February 13, 2006, lecture "A Thousand Years in Bali," delivered at the Long Now Foundation in San Francisco. Along with his outstanding 2007 book *Priests and Programmers: Technologies of Power in the Engineered Landscape of Bali* (Princeton University Press), I found Lansing's 1993 paper (with James Kremer) titled "Emergent Properties of Balinese Water Temple Networks: Coadaptation on a Rugged Fitness Landscape" (*American Anthropologist* 95, no. 1: 97–114) and his 1987 paper "Balinese 'Water Temples' and the Management of Irrigation" (*American Anthropologist* 89, no. 2: 326–41) valuable.

For related readings on technology and its healthy and unhealthy impacts on society, I suggest Edward Tenner's *Why Things Bite Back: Technology and the Revenge of Unintended Consequences* (Vintage, 1996), as well as Leo Marx's *The Machine in the Garden: Technology and the Pastoral Ideal in America* (Oxford University Press, 1964). For readings on the efficiency trap, see David Owen's *The Conundrum* (Riverhead Books, 2012) and Steve Hallett's *The Efficiency Trap: Finding a Better Way to Achieve a Sustainable Energy Future* (Prometheus, 2013).

For a discussion of how military technologies have fueled social technologies, see Michael White's 2005 book *The Fruits of War: How Military Conflict Accelerates Technology* (Simon & Schuster). On the designer and intentional fallacies, see Don Ihde's chapter "The Designer Fallacy and Technological Imagination" in the 2009 book *Philosophy and Design: From Engineering to Architecture*, edited by P. Kroes et al. (Springer).

Levent Orman's quote is from his 2013 article "Technology and Risk" (*IEEE Technology and Society Magazine*, Summer 2013, 26). Diego Gambetta and Steffen Hertog's quote is from their 2009 article "Why Are There So Many Engineers among Islamic Radicals?" (*European Journal of Sociology/Archives Européennes de Sociologie* 50, no. 2: 201–30). Robert Fein's quote on mental illness is from his presentation "National Security: Assassination, Interrogation, and School Shootings" at the National Academies' "Social and Behavioral Sciences in Action" symposium on September 24, 2012.

* * * *

Richard Nelson's 1977 book *The Moon and the Ghetto: An Essay on Policy Analysis* (W. W. Norton) and 2011 paper "The Moon and the Ghetto Revisited" (*Science and Public Policy* 38, no. 9: 681–90) are terrific scholarly resources.

Another good source is Nobel laureate Herbert Simon's 1973 piece "The Structure of Ill-Structured Problems" (*Artificial Intelligence* 4: 181–201). For more on social costs, see Nobel laureate Ronald Coase's paper "The Problem of Social Cost" (*Journal of Law and Economics*, October 1960, 1–44).

In his 1968 paper by the same name, biologist Garrett Hardin highlights his concept "the tragedy of the commons" (*Science* 162 [December 13]: 1243–48)—"The population problem cannot be solved in a technical way, any more than can the problem of winning the game of tick-tack-toe"—confirming the idea that engineering-inspired technology alone is not sufficient, but engineering-informed strategy will be critical.

Fade-Out: A Mind-set for the Multitudes

On his engineering background, Alfred Hitchcock said, "After I left the Jesuits, I went to a school of engineering and navigation, studying engineering, electricity, mechanics, the laws of force and motion, and draftsmanship. I had to learn screw-cutting and black-smithing, work on a mechanical lathe, the whole works. One got a thorough grounding there." This quote is from the 2003 book *Alfred Hitchcock Interviews*, edited by Sidney Gottlieb (University Press of Mississippi), page 164.

Popular-culture essayist Chuck Klosterman offers a crisp characterization of Hitchcock's approach in his 2004 book *Sex, Drugs, and Cocoa Puffs: A Low Culture Manifesto* (Scribner): "Alfred Hitchcock's success as a filmmaker was that he didn't draw characters as much as he drew character *types*; this is how he normalized the cinematic experience" (page 31).

Hitchcock's systematic notes, including his 1962 "Background Sounds for the 'The Birds,'" in the collections of the Margaret Herrick Library, Academy of Motion Picture Arts and Sciences; Tony Lee Moral's 2013 book *The Making of Hitchcock's The Birds* (Kamera Books); and Kyle Counts and Steve Rubin's article "The Making of Alfred Hitchcock's The Birds" (*Cinefantastique*, Fall 1980) were very helpful resources.

The articles on bird attacks that inspired Hitchcock are from the *New York Times* ("Man in Bush Wants Birds Kept in Hand," May 16, 1961); *Los Angeles Examiner* (Frank Lee Donoghue, "Man's Face Badly Slashed by Owl," May 24, 1961); *Herald Express* ("Birds Block Traffic: Invading Flocks Jam Santa Cruz," August 18, 1961); *San Jose Mercury* ("City Deep in Feathers," August 19, 1961); and *Los Angeles Herald* ("Dying Birds Jam Santa Cruz," August 18, 1961), from which the quote "The place was black with them" was taken. Reference to the 16-millimeter films is in a letter from Suzanne Gauthier (Hitchcock's assistant) to Joseph Lateana, August 23, 1961. Daphne du Maurier's piece is "The Birds" (*Good Housekeeping*, October 1952, 54–55, 110–32).

Except as noted here, Hitchcock's quotes were accessed on YouTube or at http://www.hitchcockwiki.com. "Looking at a nightmare" is from a 1965 interview in which Hitchcock discusses the crop duster scene in *North by Northwest* (accessible on YouTube). The phrase "chilling movie audiences long before air conditioning" is from a 1963 Universal-International Newsreel, "Suspense Story: National Press Club Hears Hitchcock," with voiceover by Ed Herlihy. "Dipping their toes in the cold waters of fear" is from a Hitchcock interview on the *Dick Cavett Show*, filmed in 1972, which is available as a summary by Lorraine LoBianco entitled "The Dick Cavett Show: Alfred Hitchcock" on the Turner Classic Movies website. The quotes "a man in one position for the whole picture . . ." and "I make a film entirely on paper . . ." are from Tim Hunter's interview article entitled "Alfred Hitchcock at Harvard" (*Harvard Crimson*, October 14, 1966). "Cinematically, there wasn't a single shot . . ." is from an interview between Alfred Hitchcock and Keith Berwick, for a Channel 28 program called *Speculation*, broadcast in 1969.

"You could not take the camera . . ." is from a 1964 interview with the Canadian Broadcasting Corporation, directed by Fletcher Markel (*Telescope*). "As the film went on there was less and less violence . . ." is from a July 5, 1964, interview between Hitchcock and Huw Wheldon, filmed for the BBC television program *Monitor* (also included in Gottlieb's *Alfred Hitchcock Interviews*, page 69). The quotes "extract a little more drama out of ordinary sounds" and "of course, I'm going to take the dramatic license . . ." are by Hitchcock in conversation with François Truffaut (Part 24) in August 1962—when *The Birds* was in postproduction. "Prodigious . . . I mean films like *Ben Hur* . . ." comes from the January 1963 issue of *Movie* magazine (with Ian Cameron and V. F. Perkins), pages 4–6, reprinted in Gottlieb's book, page 45. "I usually wear a blue suit . . ." is from a video clip containing excerpts of Hitchcock's interviews. The quote "Some directors film slices of life . . ." is from François Truffaut's 1985 book *Hitchcock* (Touchstone, revised edition), page 103. "You see, I like you . . ." is from a video titled "How Hitchcock Got People to See 'Psycho,'" accessed on the Oscars website.

The bird food information and "toneless composition" are from Sam Lucchese, "Birds Steal Show in New Thriller" (*Atlanta Journal*, April 12, 1963). Jimmy Stewart discusses his work with Alfred Hitchcock in a 1984 interview on the French TV show *Cinema*

Cinemas. Stewart's quote "You sort of get a feeling that Hitch's life . . ." is from his speech at the March 7, 1979, American Film Institute ceremony honoring Hitchcock. Gary Rydstrom's quotes are from the video "An Analysis of Alfred Hitchcock's Use of Sound" accessed at the Audio Spotlight website. Jack Sullivan's 2006 book *Hitchcock's Music* (Yale University Press) is a good related resource.

Hitchcock's other notable design accomplishments include the uses of kinetic typography in *North by Northwest*, the whirlpool dolly—zooming forward while dollying backward—for *Vertigo*, and subdued and extremely effective 3-D in *Dial M for Murder*. Hitchcock never won a competitive Oscar, but he did receive the Irving G. Thalberg Memorial Award at the Fortieth Annual Academy Awards in 1968.

Smithsonian historian Tom Crouch's quote is from the PBS *Nova* article "The Unlikely Inventors" (November 11, 2003).

ACKNOWLEDGMENTS

This book emerged somewhat as an assignment from Charles Vest during an elevator ride. Chuck was one of the most admired leaders in the realm of engineering education and policy. He was always a humble lad from West Virginia, even after becoming the president of MIT and, subsequently, the National Academy of Engineering. I was fortunate to have Chuck as a mentor who encouraged me to go on a "vision quest" for this project. He also inaugurated my research for this book by granting me the first interview. Sadly, he passed away about a month before I could send him my first draft. This work conveys my obeisance to Chuck.

I appreciate the support of my distinguished colleagues at the National Academies, particularly Kevin Finneran, who has been influential in shaping my understanding of several big topics, and Steve Merrill, who took a chance and opened the doors for me to the world of public policy. Hearty thanks to Randy Atkins, Clyde

Behney, David Butler, Patrick Kelley, Rose Martinez, Michael McGinnis, and Proctor Reid for their guidance and input.

Harvey Fineberg was very generous to me with his advice and mentorship, despite his ultrabusy schedule as the president of the Institute of Medicine, and so was George Whitesides with his ebullience and brilliance. Norm Augustine, Ruth David, and Charles Phelps are genuine systems thinkers, and they have helped chisel my thinking (and manuscript draft) with their acumens. For their sagacious insights, I am thankful to Paul Citron, Rita Colwell, Victor Dzau, Lonnie King, Tracy Lieu, Howard Markel, Bill Ostendorff, Rino Rappuoli, and Ted Shortliffe. They have helped fortify my professional worldview.

I worked on the bulk of this book during the dawns, dusks, and weekends of the two consecutive years that the National Academy of Sciences celebrated its 150th anniversary and the National Academy of Engineering, its 50th anniversary. I thank the presidents—Ralph Cicerone and Dan Mote, both engineers—for their encouragement in an extraordinary research setting that has been my intellectual home.

*   *   *   *

Presidential inauguration poet and civil engineer Richard Blanco once said that language is "engineered, like everything else." This book is also a product of prototyping. The lead person to have steadfastly championed my efforts is my agent, Michelle Tessler. She routinely tolerated my dumb ideas and dreadful drafts with such grace and nobility, and she placed me in the orbit of my esteemed editor, Brendan Curry, my guru and coengineer in this creation. This book and I are enormous beneficiaries of his prescience and perspicacity. I appreciate the support of his colleagues, including Sophie Duvernoy, Mitchell Kohles, Nancy Palmquist, and Alice Rha at W. W. Norton & Company. I applaud Stephanie Hiebert's superlative work with copyediting and fact-checking.

At Penguin Random House India, I am grateful to Udayan Mitra and Anish Chandy for their psychological boosts and vital suggestions that, coupled with Chiki Sarkar's encouragement and Caroline Newbury's support, have tremendously benefited this project. I appreciate the efforts of Andrew Gordon at the David Higham Agency, and Mike Harpley, Alex Christofi, Jonathan Bentley-Smith, and Lamorna Elmer at Oneworld in London, and Fuhua Ling for Citic China. I thank Michelle's interns James Barraclough and Makenna Elizabeth Sidle from New York University, and Frank Anderson from Rutgers, who provided useful suggestions on my earlier drafts.

❖   ❖   ❖   ❖

My visionary friend and invaluable guide Barbara Oakley once told me that a "good book generally has a good author—and a supportive village of friends." Prominent in my village is Toinette Lippe, who has periodically uplifted me through her mind-expanding prudence and exquisite Chinese brush paintings. Sarah and Richard Gueldner have enriched my life through their generosity and inestimable support. Heather MacAndrew and David Springbett have taught me the basics of creative thinking through their documentaries.

For their cordiality and fellowship, big cheers to Maria Dahlberg, David Dierkesheide, Muthu Krishnan, Scott Levin, Derrick Martin, Anne-Marie Mazza, David Proctor, K. P. K. Rajarajan, Kinpritma Sangha, Lauren Shern, Kathleen Stratton, Rachel Taylor, Mary Thomas, Joel Wu, and Rieko Yajima. Special thanks to Claudia Grossmann for numerous stimulating conversations in the eateries of Penn Quarter, and Adam Winkleman for robust thinking sessions, especially at the Amsterdam Falafelshop and Tryst in Adams Morgan.

My longtime friends Joshua Brandoff, Jason Cole, and Jeffrey Peake endured brain hiccups from my rough ideas and ramblings, and they saliently helped improve this book's contents. I am immensely thankful to Lindsay Bedard, Zachary Pirtle, Robert Pool, and Aparna Subramaniam for their astute and detailed comments on draft versions. For additional critical readings, my big thanks go to Subbiah Arunachalam, Susan Barker, Luke Bisby, John Caminiti, Deborah Finkelstein, and Brett Goldberg.

The works and leadership of Bruce Alberts, T. K. Partha Sarathy, S. Sathikh, M. S. Swaminathan, and David Sloan Wilson, among many others, have had important influences on my engineering education. IEEE—the world's largest and preeminent professional society for the advancement of engineering and technology—has been pivotal in guiding me through my failures, and providing valuable life lessons and growth opportunities. I recognize my colleagues Chris Brantley, Jonathan Chew, Scott Grayson, Russ Harrison, Erin Hogbin, Vishnu Pandey, John Paserba, Ed Perkins, Barry Shoop, Nicole Skarke, Leo Szeto, Rob Vice, and Jim Watson.

◆   ◆   ◆   ◆

No one has influenced me as profoundly as Dennis Hartel. He is my personal hero and has my limitless admiration. I am indebted to the wisdom of my friend Geeta Bhatt, as well as the altruism of Brian Alle; Virat Bhatt; Arthur Coucouvitis; Barbara Croissant; Padmani Dhar; Ramesh Dorairaj; Peter Fell; Eric Ford; James Hartel; Darlene Karamanos; Nong Louie; Margo Martin; Christopher Quilkey; Mohan and Savithri Ramaswami; David and Janet Rubenstein; Evelyn and Paul Saphier; Comal Subramaniam; Varsha, Varun, and Raja Subramaniam; Ran Suzuki;

Asha Unni; Thinium Vaidyanathan; Swami Venkatraman; Martin Wolff; and Patricia Zarraga.

My parents and grandparents have my unbounded veneration for their sacrifices and support. The rest of my family members have my profuse thanks for their unconditional backing. My utmost gratitude goes to my wife, Ramya. Her kindness, friendship, and commitment to greater good inspire me every day, and make it all worthwhile. The first movie we watched together was *The Birds* (a celebratory screening at BFI Southbank in London), which Ramya kept as a surprise. It was the first time I ever saw an Alfred Hitchcock film. It turned out to be an auspicious second date.

INDEX